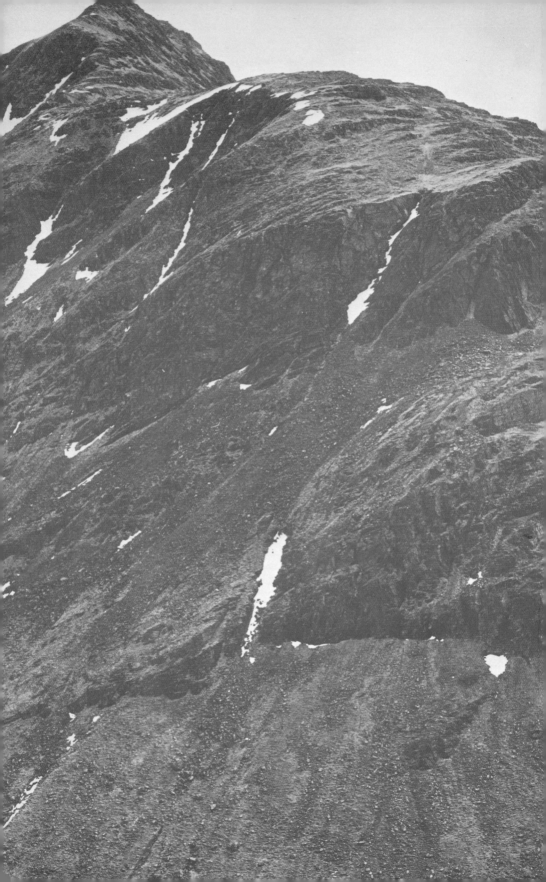

Isolation Shepherd

ISOLATION SHEPHERD

Iain R Thomson

With drawings and photographs by the author

INVERNESS 1983

ISBN 094 6597 022

© Iain R. Thomson 1983

Published by Bidean Books

Designed and prepared for publication by Ochil Books

Distributed by Ochil Books, 48 Castlelaw Crescent, Abernethy, Perthshire

Photographs by the author and Barry Knight

End paper photographs: Front — Bealach an Sgoltaidh, Back — Loch Monar from the perimeter fence of Strathmore croft

Set in Bembo by Watt Chapman, Dunfermline, Fife and printed by Butler and Tanner, Frome, Somerset

iv

To Betty

Contents

*The photographs illustrating this book
appear between pages 84 and 85*

Isolation Shepherd

The hills' wide symphony of silence
Sweeps down by a far lost way;
Music of isolation and peace
Carries time to horizon's rim
Where the wistful plovers call
Plaintive clear, on a distant day,
Reaching harmony's only source
In the spirit of a single mind.

But gone is the ear that hears it,
Lost is the breath of care;
Scattered the race to dreaming
Fond eye to Corrie and Beinn.
Heavy the loss that is Highland
For the hills of striding men.
Shelter the glens that are weeping
In the care of a single hand.

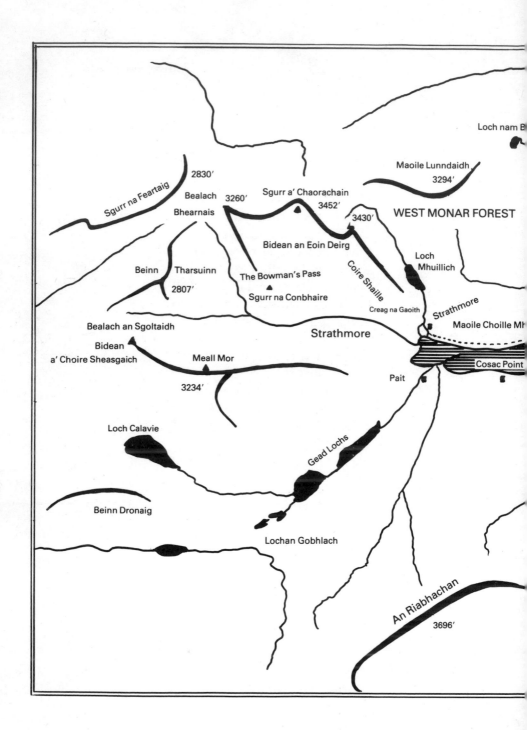

Loch nam B⋯

Maoile Lunndaidh
3294'

WEST MONAR FOREST

2830'

Sgurr na Feartaig

Bealach 3260'
Bhearnais

Sgurr a' Chaorachain
3452'

3430'

Bidean an Eoin Deirg

Coire Shaille

Loch
Mhuillich

Beinn   Tharsuinn

2807'

The Bowman's Pass

Sgurr na Conbhaire

Creag na Gaoith

Strathmore

Maoile Choille Mh⋯

Bealach an Sgoltaidh

Strathmore

Bidean
a' Choire Sheasgaich

Meall Mor

Cosac Point

3234'

Pait

Loch Calavie

Gead Lochs

Beinn Dronaig

Lochan Gobhlach

An Riabhachan

3696'

X

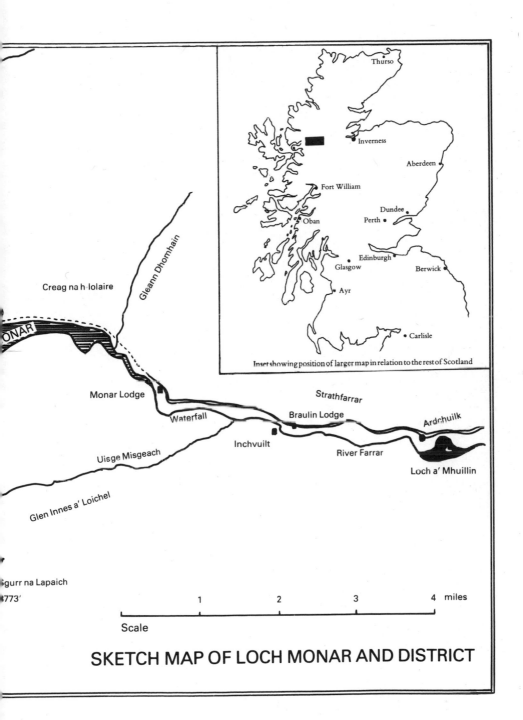

Creag na h-Iolaire

Gleann Dhomhain

ONAR

Monar Lodge

Waterfall

Uisge Misgeach

Glen Innes a' Loichel

Sgurr na Lapaich
3773'

Strathfarrar

Braulin Lodge

Inchvuilt

Ardchuilk

River Farrar

Loch a' Mhuillin

Thurso

Inverness

Aberdeen

Fort William

Dundee
Oban   Perth

Edinburgh
Glasgow   Berwick

Ayr

Carlisle

Inset showing position of larger map in relation to the rest of Scotland

| 1 | 2 | 3 | 4 miles |

Scale

SKETCH MAP OF LOCH MONAR AND DISTRICT

BY STORM TO STRATHMORE

A south westerly gale and heavy showers swept down Loch Monar.
It had been blowing and raining since the previous day. Though
summer storms are not infrequent in the high hill country of the
Highlands, this one was severe. The *Spray,* a clinker built 26 foot
ex-ship's lifeboat was to demonstrate her qualities in dealing with
rough conditions as we left the shelter of the 'narrows' at the east end
of the loch. Head on she met the full force of the weather in the wider
open waters. Her cargo that particular day, 13th August 1956, was my
family and flitting, destined for a new home, six and half miles of
stormy loch, westwards from Monar. Here, cradled in remoteness
and grandeur at the upper reaches of Glen Strathfarrar, lay
Strathmore.

We were quickly to learn that the wave action on large fresh water
lochs is quite unlike that of the sea. Wave follows wave in quick
succession, deep troughs and sharp breaking tops make dangerous
conditions especially as a boat lacks the buoyancy it would have in salt
water.

Iain MacKay leaned over the tiller. The *Spray* drove into a press of
surging water. Shielding my eyes I stared ahead. White rolling tops

stretched to the grey indistinction of storm swept hills. Astern the longstreaked wave backs heaved powerfully away from us. Capricious gusts, sometimes snatching the crumbling crests, threw spiralling sheets of water to meet the rain. Our world shrank to simple elements, raging and shrieking, warning or welcome. Gone the false world of human progress; I felt the first thrill of wild isolation.

Handling the heavily loaded launch required skill. Occasionally bursting through the crest of a viciously curling wave she would crash into the following trough with a solid thud which shuddered the whole length of her timbers. Arched sheets of water shot into the air to be caught and hurled across the huddle of us crouched at the stern.

Sometimes a broader wave would lift the stern until the propeller almost cleared the water allowing the four cylinder engine to rev and clatter alarmingly. Rising to the next wave, the propeller dug deep, biting much water. The engine dropped to a sickening struggle. With a hint of unspoken concern Iain would give her more throttle. Should the engine stall in such heavy conditions and the *Spray* turn broadside, well, we preferred not to think.

Violent gusts bore down on us, whipping rain and spume into our screwed eyes. The two MacKay brothers and myself were bent oilskinned figures in the exposed engine cockpit. Green tarpaulins running with water covered our worldly belongings in the centre well of the boat. Across them I glanced at the family. They sat apprehensively under the open fronted hood which served as a cabin two thirds of the way for'ard. I had visited Monar some weeks previously when engaged as Shepherd but this was my family's initiation to a wind tossed lonely world. Perhaps the more fearsome for Betty as she was unable to swim. To my relief I saw that under a shawl she was quietly feeding Hector our ten day old baby. Alison, his two year old sister sat close to her mother, wide-eyed but silent.

The passage west to Strathmore, due to the conditions that day took about an hour and a half, whereas a trip up the loch in fair weather could be done by the *Spray* in forty minutes. Little was to be seen of the majestic hills, only fleeting glimpses of black and green betokened their massive presence as mist and cloud, swirling low before the westerly blast, clawed across their aristocratic faces.

The loch was in high flood. This added power to the waves as we surged westwards towards a line of breakers stretching across the head of these open waters. I could see the reason for these rough conditions was a sandbank, a natural formation which reached out from the south side of the loch in a sweeping curve to within forty yards of the north shore. A spit of sparkling mica sand, it generally showed well clear of the water level. On normal days a feat of navigation was required to negotiate this narrow channel, or The Corran as we called it. A dog legged swing past an iron standard marker veered one's boat hard towards the north bank before a smart turn avoided apparent disaster and the tricky passage afforded access to the head of the loch.

Young Kenny MacKay took the handkerchief from about his throat, "we'll not be needing to use the Corran today," he said wiping his face. "The level must be eight feet up and the half of it feels down my neck."

Trusting to their judgement the boys stood the boat into the heavy breakers. A moment hung tense. Would we strike? Twenty yards, spray and violent motion: I saw the boys relax, we were across the shallows.

Astern of us I spotted the Monar launch making up at speed. She sliced into the waves throwing water aside in fine 'clipper' style. A fast sound boat, again clinker built, but lacking the beam of the *Spray* she was not so suitable for cargo. Allan Fleming the Monar keeper, Mr. Roderick Stirling my new employer and 'Big Bob' Cameron the ghillie were aboard her. Their deerstalker bonnets bobbed above the cabin roof as they too judged the depth of the sand bank before sailing over it.

As Iain slowed the *Spray* as much as he dared, I looked about me. To our north the few buildings at Strathmore came into view. The usual landing, a wooden pier and zinc roofed boathouse normally at the waters edge, stood awash, waves at its rafters. The small intimate stone built shooting lodge, surrounded with birch and pine was reached by a path leading up from the pier. It looked dank and empty. Some few hundred yards further along the pathway, beyond the swaying lodge plantation, stood the tiny shepherd's house. Benefiting from a few faithful birch trees it was still rather exposed being built within a croft of a few acres and away from the old homestead. Sheep fanks, barn and byre were nearby, stable and bothy perched on a pine clad knoll overlooking the croftland. It became our home for the next four years.

Betty looked out uneasily at the sparse scene, glimpsing through still driving squalls the steep threatening hills that leaned imposingly over Strathmore. Heightened by scudding mists, dark and uninviting, they jeered down a challenge to our intrusion.

The Monar launch drew alongside, bobbing and heaving the boats lay abeam. Conversation crossed in hand cupped shouts.

"Not a chance of the pier today boys, what about trying the lochan below the house" young Kenny yelled against the noise of gale and engine. Only catching the odd word, "Head up below the house," Allan bawled back moving off his launch to feel his way into the channel leading to the 'wee loch.' Round from the boathouse I saw that the flood stretched through croft and fences almost to the house. In dry conditions a channel did exist but could only be entered with difficulty in a rowing boat. There was no such obvious passage now, just a sheet of tossing water.

Ahead we watched the Monar boat enter the bay without grounding and run her bow hard in on the grassy banks only eighty yards below the cottage. Big Bob leapt ashore with a rope and made secure. In the more heavily laden boat we moved with extreme caution, submerged rocks being Iain's concern. Edging in with steerage, no more, all well, then a slight lurch and hesitation. We looked sharply at each other, "hidden peat hag?" A few more yards, we waited. Too late, with a sudden heel she grounded.

In the mischievious way of the weather the gale found fresh vigour. Immediately the *Spray* began to swing across the channel. Iain spun the engine into reverse. Following Kenny I sprang out of the cockpit. Balancing up the deck we grabbed oar and boathook from the hold. Betty looked alarmed. Poles down into the mud we heaved and strained to help the engine. For minutes nothing happened, a heavier wave passed under us, we swung broadside to the channel, then mercifully, slowly, almost reluctantly, she slid off the bank. Both of us were now to one side pushing the bow back on course. At engine and tiller, Iain set us moving again towards the group now ashore.

Twenty yards out from the water's edge, and without warning we struck again. Drawing so much water this would be as close in as we could get. Turning slowly we lay wallowing in each wave. The unrelenting rain hissed on the water. Nothing for it, without word or hesitation, Kenny went over the side. Chest deep he waded out with the bow rope. Taking his example I plunged in with a stern rope and together we secured the boat. My family peered out from the shelter of the hood, doubtful of their fate. Iain laughed, "This is the terminus, tickets please."

Joining the Monar group now almost as wet as Kenny and myself, this was not a day for 'newsing'. Allan hid a smile as water squirted out of our wellingtons, "Ah well, you're getting wet boys," he sympathised with a grave tone, going on to explain, "Bob's away for the cartie, not a day for the pony, we'll just make do with the big chap."

Big Bob wasted no time and appeared trotting down the croft, his lanky form between the 'trams' of a rubber wheeled horse cart. "I'd rather be a pony here than a ghillie," he panted, obviously alluding to 'Dandy', who doubtless, was knowingly snug in the Pait stable, unable to be swum over the sandbank, thanks to the flood. Bob, a tall, broad shouldered man was singularly powerful, nor did he spare himself. Keeping the cart's back chain over his shoulders he reversed it down into the water towards the side of the *Spray*. In the last few yards the light contraption began to float. With blunt commands generally reserved for intractable ponies, we ordered Bob to stand and hold the cart steady. It rose and fell gently in the lea of the boat.

'Birkenhead' fashion, we decided women and children first. Iain and I stood on the 'cartie' it slowly sank until its wheel touched the bottom. Betty handed a warmly wrapped sleeping baby over the side to me and urged Alison to jump to Iain. She then stepped gingerly down onto the submerged cart, the cold water rose unpleasantly to her knees. We kept a swaying balance, the boys pulled heartily and the cart reared out of the shallowing water. In a minute the family stepped down at the foot of the croft. Betty looked immediately more cheerful and taking baby Hector from me set off up to the house. Alison, always a lively child whatever the conditions, skipped alongside. At last the rain was easing and a welcoming coil of smoke rose from the cottage chimney.

For us, it was back into the water to commence the final stage of the 'flitting'. Peeling back the tarpaulins I saw much of the furniture seemed in a sodden state. "Don't worry Thomson, it can only get drier," they assured me as we abandoned any pretence of defying the moderating weather. Still working waist deep we loaded the cart and trundled the soggy 'valuables' up to the house.

Our most difficult item was the piano. A heavy old brute of an instrument which we transferred from boat to cart with much inelegant comment and no little sweat. Lying on its back in undignified fashion, the lid fell gaping open as we hauled it up on the penultimate load. I viewed with aesthetic concern its trembling innards. Would it ever play a note again? "Won't be too keen on Handel's Water Music after this wee outing," observed a wit, displaying a hitherto unsuspected depth of musical insight. I had to admit, it did appear a quaint mode of delivery for an instrument of such obvious quality.

The tiny kitchen at Strathmore was indeed a cheerful refuge that day. As I entered with our final load of furniture I found Mrs. MacKay and her red haired daughter Biddy busying themselves before a roaring stove. In spite of the weather, they had rowed over that morning from Pait with supplies and utensils to scrub out the house. It was early evening as we gratefully drank the soup and tea they had kindly prepared. Dripping oilskins created large pools around each pair of wellingtons as 'news' of the glen began to circulate.

In the nature of summer storms the evening brought a clearing. Clouds were lifting from the hills in trails of torn vapour as we looked out of the kitchen window to where the boats lay beached. The MacKay boys were anxious to get away before the loch began to fall and strand the boats. Mrs MacKay and Biddy clambered onto the cart to keep their feet dry, one more wetting for the men and, all aboard, the boats pulled away. The MacKays back to Pait pier and their home a good mile across from us on the south side of the loch. As I watched them go, a spotlight of sun through parting clouds picked out their cluster of buildings and trees, an oasis of life amidst a vast sweep of lonely hills. Allan the keeper, with a cheerful wave, headed away down to Monar. The swell at their backs would make for a fast run.

Suddenly Strathmore became our own. I felt its emptiness, its pristine grandeur, and, still faint, its irresistible call. With a clear note of warning and censure however, the doctor who cared for the baby's confinement had commented "you're not taking a week old baby into that wilderness, no road, no telephone, dozens of miles from anywhere, it is not being responsible". I walked back to the house his words in my ears. The skies were brightening, hidden sunlight fringed the edge of towering clouds. Away to the west, in the fine steep sided glen of Strathmore, the hills took shape, their green summer slopes streaked with veins of white foam. The roar of the Strath Mhuilich burn from its spidery waterfalls just to the east of the house came and went as the wind eased.

The cottage, small but comfortable, had two downstairs rooms and bathroom. Upstairs, the two attic bedrooms were lit only by skylight windows. A low eaved steep roofed habitation dwarfed by the overhanging crags of Creag na Gaoith whose scarred and broken pillers tuned each passing breeze. How rightly it was named 'Rock of the Wind'.

A sky window opened over the shoulder of far Beinn Tharsuinn. The late evening sun yellowed a glen, desolate, sometimes daunting, to many, even cruel. Yet for those prepared to live simply, intimately in trust with these testing elements, might not such country slowly yield of forgotten pathways? The sun lost its last rays in the blood of life. I went inside. The bedding steamed before the fire. Alison was alreadly asleep in her cot. Betty was nursing the baby, she looked up. I knew then, all it is given to know. That night we unrolled the mattress on the floor and lay listening. Only the note of the burn and the catching voice of the wind on Creag na Gaoith sang the last of a summer's storm.

THE GREAT STRATH

GLEN STRATHFARRAR as we knew it was one of the last great unspoiled areas of the Highlands. Some thirty miles long, broad and majestic it is dominated from many aspects by Sgur na Lapaich (3773'). This beautiful symmetrically shaped hill situated towards the southwest end of the Strath carries snow well into July on the north corries and is suitably crowned by a sizeable outcrop of white quartz rock. We viewed the fine hill's westerly side which was a feature from our tiny windows at Strathmore. It constantly changed according to light and weather, and a glance across at breakfast time was something akin to a tap on the barometer.

The Farrar is a main tributary of the Beauly river, the confluence being at Erchless in Strathglass. Here stands a 13th century castle, once seat of the The Chisholm and guarding the gateway to Strathfarrar. The name Farrar goes back to Roman times when under Agricola the whole valley and estuary was recorded by him and appears notably in the pages of Ptolemy as 'Varar'. Certainly the name precedes the arrival of Gaelic in the Eastern Highlands, for here in Roman times the ancient Pictish language was spoken (a form of Celtic akin to Welsh). An older dialect still, belonging to the pre-celtic inhabitants, may just have lingered on. However if the word is Pictish, then the root 'Var' meaning to turn or wind gives a likely explanation, for the winding river is a most suitable description as its tree clad banks know many bends and with this possible deduction we must be content.

Not inappropriately at the foot of this formidable Strath was a hostelry, the Struy Inn. In our day it was the village dance hall and many a good 'fling' we enjoyed. Now it has been modernised and is the stylish residence of a farm manager.

In its heyday, the 1860's, the Inn was kept by one Roderick Urquhart, a quaint character with a flair for both old anecdote and hot gossip. Unfortunately he was no judge of whisky and those not inured to his particular brand generally suffered the pangs and penalties consequent upon indulging too freely in the seductive fluid.

At an earlier date however the Inn gained some notoriety from the murder of Maurice MacRae (Muireach Fial) a Kintail man who had grazing rights in Glen Affric. One particular evening he was returning from Inverness by way of Strathglass to his native Inverinate and chanced to call at Struy Inn. There as he sat imbibing with the Strathglass men they quickly discovered he was carrying a great deal of money. This proved his great misfortune. He was set upon, dirked to death and stripped of money and clothing. Next morning however his pony was let out of the stable and in the nature of many faithful animals quickly made home to Kintail where the men were surprised to see its return without a master.

Suspecting all was not well, they sent out a party to search the route. On reaching Struy Inn, and listening at the window, they heard the Strathglass men laughing and joking about the murdered MacRae "he looked like a white salmon bobbing about in the water" said one to loud guffaws. Down to the river went the Kintail men, where sure enough the body was found caught in the alder bushes at the junction of the Farrar and Glass rivers. They pulled it out, hid it in some bushes and hurried back to Kintail. Gathering together a large party of MacRaes they returned through Glen Affric to Strathglass, quietly retrieved the swollen body and set off home.

Passing the burying ground of Clachan Comar near the present village of Cannich they found a funeral in progress. Surrounding the kirkyard walls the aggrieved clansmen fired shots over the heads of the mourners who fled in terror. Somewhat out of character for MacRaes, on this expedition they were not out for bloodshed and contented themselves with taking away a flat stone which was intended to cover the interment which they had peremptorily interrupted. Procuring, if not, stealing a small cart they took the stone and their dead friend back to Kintail where he was buried in Clachan Duich, that beautiful burying ground on the shores of Loch Duich. You may still see the grave surmounted by a stone intended to honour the earthly remains of a man from Strathglass. Such were the violent and unscrupulous days of yesteryear. The local characters of today though colourful barely resort to such excesses.

Leaving our once homely Inn the valley widens to give the sound arable lands of Culligran, and the eye is drawn to the first of a long range of splended hills. Here high above the flats and fields of Culligran at the foot of Strathfarrar rises Beinn a' Bha'ach (2826'), its grazings being particularly lush and verdant near the summit due to a number of springs, no doubt of a mineral nature. The homestead was tenanted in the 1880's by a family named Tait who in addition to running the farm were 'posties' for the glen. The mail and supply carrying was done by horse and cart, a journey of some fifteen miles from Struy to Monar. During our days in the glen a three times a week mail run was faithfully carried out by retired wood cutter Adam Cossor whose first conveyance was a motor bike and side car, and latterly an old Morris Ten. Such was the state and surface of the road in those days, it could only really be described as a rough track involving many tortuous dips and bends. Cossor smoking his thick black twist, would tootle up the Strath summer and winter. The journey could take at least two hours to the foot of Loch Monar where our official postman, old Kenny MacKay from Pait might be waiting with the launch. Weather permitting he would convey his own and our mails by water for the last six miles taking care of course as he often joked, not to open the Pait delivery until safely at home. Although his excuse was he had signed the Official Secrets Act, privately we guessed that Teenie, his wife, insisted on first readings.

A mile west of Culligran the river Farrar passes over a fall not high but of singular force. This is due to the narrowness of the rock cutting through which the water pours. It also turns at a sharp angle and in times of spate provides an awesome sight. It faces the salmon with a real test of strength, and watching their repeated attempts was most exciting. About these falls the glen is still heavily wooded. The south side grows one of the finest and most extensive stands of old original Caledonian pine forests found in the Highlands. Rugged and inaccessible with a dense undergrowth of deep heather and bog myrtle, it was not suitable for sheep, but provided excellent deer wintering ground. Heavy snows saw large herds down amongst the welcome shelter.

The wild life in this unfrequented area of the glen is exceptionally varied. Each stretch of the river had its pair of sandpipers. Their incessant calling was in my ears early one summer morning. I was travelling west by bicycle, a mode of transport which involved as much walking as riding due to ups and downs of the road. As I skidded round a bend a wild cat half carrying, half dragging, a large rabbit crossed in front of me. It had come onto the road some six or eight yards beyond the turn and paused momentarily to look my way. As I shot towards beast and prey the ears went down quite flat and in a crouching run it made for the undergrowth below the road. From width of head and general size a male cat seemed probable. Would it be carrying to a den? I didn't stop or make it my business to find out. Meetings with various animals was a common happening and seemed in the natural order of events.

For most of his life Donnie MacGillivray keepered the Struy Estate whose land forms this part of the glen. He too used the bicycle, but as Donnie said ruefully "the road was kept up with a squad of roadmen in my day". During the 1920's he aided the work of a Capt. Knight, an early ornithologist. The Captain was a keen observer and photographer of golden eagles. There were tree and cliff eyries in the 'big Caledonian wood' and with Donnie's help some excellent early work was achieved. In the Middle Ages the golden eagle had been known as "Hawk for the Emperor". Taking this as a challenge, using skill and patience, Capt. Knight actually succeeded in training a female eagle for falconry. Eventually regarding the bird as too heavy and dangerous the Captain released her in the glen. She sailed away and was not seen again. Donnie, once entrusted with feeding her, was surprised how little she ate. "Perhaps the village children who gathered round each day made her shy" he told me with a smile.

The north side of the glen in these lower stretches is mainly birch and rowan with many jumbles of moss covered rocks. One particular rocky overhang, some twenty yards off the old road, was known as "Prince Charlie's Cave". In his wanderings the Prince certainly was in Strathfarrar and there is no real reason to doubt the story. After passing the royal cave the road cut through a narrow defile and swept down a steep hill. The effect on a sharp day was that of stepping through a doorway into the magnificence of the great glen itself. Sgur na Lapaich burst into view, the beauty of its shapely peak, though twelve miles distant, would seem to rest amongst the dark green tops of the nearby pines. This fine hill often deep blue against the sky, carries its snow fields into the warmth of July. Many a morning they sparkled in the sun-light to greet an early traveller and cheer his heart.

Reaching Deanie Lodge the Strath broadens and becomes more fertile with considerable areas of arable ground lying around the shores of Loch Beannacharan. Above these flats and to the north side is the great bowl of Corrie Deanie. From this basin by way of a long ridge walk there begins a line of extensive peaks. Many are well above 3000' and flank the glen right through to the watershed west of Strathmore. Looking down on Deanie and Loch Beannacharan at the start of this range is Sgur na Corrie Ghlais, 3554'. It presides with solid bulk over a scene of classical perspective as hills range away to the horizon. At Deanie we are now on the Lovat lands. These extensive deer forests form part of the inheritances of Simon Fraser, Lord Lovat of World War II fame and bravery. It was under his management that sheep were introduced to this part of his estates in the early 1950's. Blackfaced ewe hoggs were bought in Lanark and under the management of Jim MacLean, a Kintyre man, a large sheep enterprise was built up. Integrated with other areas of the estate as many as ten shepherds were employed. The Strathfarrar sheep stock built up to 1,600 ewes and 400 hoggs, which ran for about fifteen years before giving place once again to the red deer. Together with the sheep grazed a large herd of hill cattle. They in fact are still summered on the glen flats. The present Lord Lovat pioneered hill cattle 'ranching' after the war. He started with traditional Highland cattle but later adapted to more productive crosses, notably for many years the Irish Blue Grey. I might say that photographs of the Strath and the Lovat cattle herd which appeared in the farming press when I was a boy fired my imagination. I vowed when I saw them "this is where I shall go one day" and so it turned out to be!

Willie John MacRae was for many years the shepherd living in Deanie. The misfortune of a lame hip did not prevent Willie from doing a full day on the hill. His dogs however were of exceptional ability and must have saved him many a mile. Old Duncan Fraser neighboured him in the small house above the Lodge. Duncan, though small, was one of the hardiest of men. A stalker with the Lovat Estate in various parts of the glen all his days he survived to a good age on a mixture of whisky and thick black twist tobacco; the latter he would smoke as he climbed smartly out to the hill. Anybody following his rapid step would be enveloped in its thick cloud which would leave them coughing and choking. When well into his seventies Duncan was one day rolling wool for us at a Braulin clipping. He would do the work of two men and admiring his energy and wit I enquired of him had he any regrets in his long life. "Yes", he replied in his sharp voice "I should have drunk more whisky when it was cheap". Duncan it could be said always enjoyed a good night out in traditional style and was known to be the best of amusing company.

The Caledonian forests on the north side of Loch Beannacharan and again further west at Braulin were cut just prior to the second world war, some few trees were left here and there giving some indication of a former beauty, open and spacious in the nature of these woods. The carving glaciers of 20,000 years ago encountered granite just to the west of the Loch Beannacharan and a high outcrop of this hard and durable rock remains well above the valley bottom level. Standing like a sentinel on top of this rock bastion was a single Scots pine of considerable strength and girth. In certain light or against a mist swirling background it had an ethereal quality and made the glen seem truly primitive, a lonely lost figure watching from the hermitage of a forgotten era.

Perhaps the bonniest part of the Glen Strathfarrar is to be found at Loch Mhuillidh. This peaceful lochan contains a heavily wooded island. Scots pine, sycamore, birch and rowan often find their different shades reflected in its surface, for being relatively sheltered the loch has many calm days. Here the majesty and elegance of Sgur na Lapaich seems to bend over the scene with the lesser hills to either side accentuating this splended mountain's graceful form, a Highland view of much beauty and natural balance is presented.

The islet, Eilean Mhuillidh, profusely overgrown is not visited by deer or sheep and contains the ruins of a building. Simon Fraser, Lovat of the Forty-five, known less respectfully as 'The Red Fox' was a man of considerable ability. Amongst the intrigues and plottings rife in Jacobite affairs his pathways are difficult to follow. Suffice it to say however, that his hold over the Highlanders was extraordinary. To such extent was his standing in the north that he was, at one point late in the 1745 campaign, offered Command of the Jacobite army. A sitter on fences as suited his nature he finally, though by then old and infirm, sided openly with 'The Cause', but by which time all was not well with Jacobite affairs. He was however, with good reason, a believer in the efficacy of islets as places of refuge, for in one of his earlier escapades connected with a forced marriage, he fled the arm of the law to Eilean Aigas, his property in the Beauly river. Realising in 1746 that matters were looking dangerous he had a hideaway prepared for himself on the little Eilean Mhuillidh to which he was duly carried. As the situation deteriorated he removed to an islet on Loch Morar in western Inverness-shire where he was ultimately captured and from whence he was taken south to the Tower of London almost in the form of a road show. Beheaded on 9th April 1747 he died without fear, maintaining all the dignity and pride of race of an aristocratic Highlander.

Mhuillidh was a populous settlement at the turn of the century, old Kenny MacKay recalled having refreshments from a cailleach who was one of the last inhabitants of the clachan. When Kenny as a youth prior to the first world war passed up and down the glen, he was glad of the oatcake and whey hospitably given. Depopulation had been rapid and by 1920 the village was empty. All the dwellings were well to the top side of the road, a strategic point blissfully ignored in 1959 by the constructional engineers as they laid out the camp site which was to serve man and machine for the building of the Strathfarrar Hydro Electric scheme. Noting their on site deliberations one day, old Duncan Fraser a keeper all his days in the glen, observed "So you're planning to build on the flats below the lochan and on the low side of the road". To this questioning statement they replied that was certainly their intention. "Well, well, manys the day I saw that ground flooded from side to side" observed Duncan. The engineers smiled deprecatingly and assured him they had checked the relevant figure, made adequate calculations, and such events must be figments of an old man's imagination. A large camp was duly built that summer. To the consternation of the officials winter 1960 saw Duncan's romancing take shape. The camp was deluged under six feet of water, much damage ensued. A certain smile played on the faces of several of the locals. I must admit to being responsible for the rainfall records at the west end of the catchment area during my years in Strathmore and did not take the task too seriously, but who is now going to suggest any connection.

A mile beyond Loch Mhuillidh was the house and croft of Ardchuilk which by 1862 was the home of John Ross, head keeper, or forester as the position was termed in the eighteen hundreds to Lord Lovat. The lands about Ardchuilk, green and extensive by standards of the day sloped down to sweeping bends of the Farrar.

Making way for the deer forests in the early eighteen hundreds two families were moved from this croft land. The great grandfather of Willie John, the Deanie shepherd, was a Valentine MacRae, who with his family, farmed the lands across the river from Ardchuilk. To sweeten his flitting the then Lord Lovat called with an offer, "MacRae I will give you a place better than 'Uch a Ra' which was the name of this little holding, "No, my Lord you can't" replied MacRae, "for you haven't got it yourself". Such was the sturdy character of MacRae as the following story illustrates.

A guest of the Lovats was shooting one day in the glen. MacRae performed some service for him which included rowing the shooter across the river. As they parted the gentleman offered the Highlander a half sovereign. Politely refusing the gift MacRae told him "I have plenty money of my own". Later that night over dinner down at Beaufort the English guest recounted this somewhat patronising incident to his Lordship "Oh" laughed Lovat "that's Uch a Ra all over. I don't suppose he has two coins in his pocket but he doesn't have need of them, he's independent". Eventually MacRae along with a family of MacKenzies, who were in Ardchuilk, were removed. MacRae was offered the choice of three farms and strangely took the poorest one, which was at Carnoch in Strathglass, perhaps because it was nearest to the glen he loved.

The last keeper in Ardchuilk was one MacDonnell who brought up a family of six girls and a boy. In his youth Duncan Fraser was there as a ghillie. By my time in the glen however the old house and its environs were becoming ruinous and had been converted with a gate or two for sheep handling. Duncan told me on one occasion as we worked there dosing some hoggs, that conditions were so bad in his day on a wet night he had to put up an umbrella over his bed!

A dance was a great event amongst the population of the glen in those days and Mrs MacDonnell of Ardchuilk was obviously a lively wife. However on the night of one particular Struy dance it was starting to snow and MacDonnell a less enthusiastic waltzer than his spouse was unwilling to leave the fire. The resourceful Mrs MacDonnell instructed Duncan to go through to her husband and inform him that Peter MacDonald the head keeper from Braulin had just passed down en route to the function. Duncan duly did as he was bid and MacDonnell after some thought begrudgingly told him to put the pony in the trap. Off they set. The storm grew worse. They arrived at Struy only to find the head keeper was not there at all for he was too good a judge of the weather. MacDonnell, doubtless somewhat displeased, set off back home. West of Deanie they got stuck in mighty drifts and the toe tapping wife had to be loaded onto the pony for the remainder of the journey. It was six weeks before the trap was extracted and taken home and an equal time might have elapsed before a thaw reached the MacDonnell household.

Travellers in Strathfarrar even today will notice three lime trees beside the road at a spot about a mile west of Ardchuilk. It was the policy of Lord Lovat in those days to plant limes beside each spring or well, the theory being that the tree would in some way purify and sweeten the water. These three trees also provided shade and Lord Derby, one of the great shooting tenants used to picnic beneath their spreading boughs. The spring on the side of Loch Beannacharan was similarly provided with lime trees. It ran into a trough which the horses would never fail to make for when passing.

We have now travelled ten miles westwards up the glen. Suddenly as we cross a sharp rise in the road are spread before us the spacious alluvial Braulin flats. The Farrar here winds lazily between high sometimes crumbling banks which give plenty of evidence of deep soil. Criss-crossed with drains and ditches the area is now heavily rash infested providing nesting sites for many duck and wader. Though unhealthy for livestock they are nevertheless a sweet attraction to sheep and deer especially in the spring. I have seen these flats covered in hundreds of deer. As one disturbed them when passing along the road lying to the north edge of the flats, a fine sight they would make stringing out in lines across the river and up the south side of the glen into a vast stand of Caledonian pine. The deer unless forced hard, always crossed the river at certain fords their galloping feet making a lively splashing sound. Stately old pines on the south side give a fine open forest, stretching some five miles west to the lower slopes of Sgur na Lapaich where they gradually give way to birch. A start was made by the Canadians to extract this timber during 'Hitler's War' but it proved too difficult and was abandoned. Many trees, mature giants, lie fallen, their sound timbers so slow to rot that they provide shelter for regrowth and thus another generation arises. Most of the biggest trees here would have seen the year of Culloden.

Traversing the 'Rough Wood' as we called it, either at a fox drive or a sheep gathering with the Lovat shepherds, I was always struck by the security these trees gave. The silence and shelter they provide was somehow re-assuring when one came down from the openness of the hill. Surely there is not a more magnificent species of tree than our Caledonian pine. As they stand against the elements the strength and spread of their great roots grip the ground like the muscular veined hand of a powerful old man. Gnarled and twisted limbs emerge from a stocky trunk often of immense girth relative to the tree's height. The rich red brown bark is seen at its most appealing in the warmth of the gloaming. Soft yellow evening light slowly fades and the dark green of the needle canopies turn almost to black. The fragrance of these open woods is of special charm, pine, bog myrtle and most delicate, the wild orchis. Cutting these giants by cross-cut saw releases the full, almost intoxicating aromas of their sticky golden resin.

Braulin Lodge built in the 1840's by Lord Derby, a shooting tenant of the Lovat estates, looks across to the 'Rough Wood' from its position on the north edge of the flats. A mellow building of fine red sandstone masonry, it blends with the scene - more than can be said for the wooden two-storey bothy built alongside it. Doubtless, however the bothy could tell a tale of 'high jinks' in the old days between ghillies and maids. Invariably such matters were a cause of concern to the Lady of the House as she strove to protect the innocence of the girls under her charge. The staffing of the lodges during the stalking era brought new blood to many areas.

Braulin forest became noted for its great stags. George Ross descendant of the Ardchuilk Ross family tells of the heavy maize feeding that was once practised in order to bring out their heads. As a boy of fourteen years George came to Braulin from his parents' home beside Beaufort Castle, seat of the Lovats. He had no experience of Red Deer and stayed by himself in a wee bothy beside the Braulin stables. One night at the beginning of the rutting season, many large stags came about the buildings where they were accustomed to being fed during the winter. The lion like roaring commenced. Only fourteen and alone, George, not knowing what was making the deafening noise was petrified. He barricaded door and window, vowing that if he got out alive he would flee for home at daylight. Badly shaken, next morning he went straight to the head keeper Peter MacDonald, or "Peter the Scout" as his by-name had it and told him he was for off. MacDonald laughed till his sides ached but young George stayed and became in his turn a most successful head keeper.

Glenstrathfarrar divides at the head of the Braulin flats where the keeper's house and croft of Inchvuilt is situated. Lying just across the river it is reached by a tall wooden bridge above which lies a fine salmon pool suitably convenient to the house. Old Duncan Fraser told me that in his youth when the salmon were spawning in late summer all the hill burns that run into the Uisge Misgeach (drunken water) as this tributary of the Farrar becomes called when it reaches west into the Gleann Innis a Loichel, were packed with prime fish struggling for the best gravel beds. The twisting Uisge Misgeach runs three miles to the watershed in the Bealach below Sgur na Lapaich which is the march between the Lovat estate and Pait forest.

The Farrar itself bears northwest from this confluence and into extremely bleak, inhospitable country; dark weathered rocks are barely covered with vegetation and the ugly rocky mass of Beinn na Muice rears close to our tortuous westward track, which plunges and twists in a most alarming fashion for the remaining three miles to Monar. Just half a mile short of a deep, narrow gorge through which Loch Monar pours, powerful and spectacular falls could once be seen far below the road. These Monar falls were both wide and high, the water forming a jagged rock-strewn cataract before gushing down in several mighty shoots. In spate the falls lay hidden in a steamlike spray and the roaring sound rumbled with a thundery menace around the closing hills. The falls were sufficiently mighty to prevent salmon from reaching Loch Monar and indeed they would then have reached far beyond to the headwaters of the Farrar, only eight miles from the west coast at Loch Carron.

As the road entered this narrow gorge it became dangerous. On one occasion whilst waiting for cattle (which were being floated up from Fairburn the home estate) to arrive, we were surprised when the beasts arrived walking quietly towards us. Thoroughly alarmed we raced down the road to find the float had run back on a steep section and turned over the edge. It lay 'cowped' on its side. The cattle had walked out through the top of the float and left the dazed but lucky driver to make the best of it. The old road through this fearsome defile was even more precarious and those who chanced to take their eyes from the 'new road' could see sections of a former track perched high amongst the cliffs some six hundred feet above the river. Many coaches coming up from MacRae's, the hirers in Beauly, were driven four in hand along it. John Collie, a Monar keeper of the 1860's, remarked in his little book of memoirs, "The road is narrow in places and dangerous if your horses are driven rapidly".

Breasting the crest of this hard-won windswept pass, swing to the right and you journey in a matter of yards into a land transformed. It is perhaps the most breath catching shift of scenery in the north. Highland grandeur stretching before you in its finest form. Away to the distant blueness on a heat shimmering summer's day glitter the shining waters of Loch Monar. Tree clad tapering promontaries run out in rock and sand to its tranquil depths. The eye is drawn hungrily to the wide amphitheatre of hills, ridge upon ridge, with the tall sharp peak of Bidean an Eoin Deirg rising majestically above Strathmore and gracing the scene. 'High top of the red bird', all its splendour of steep corrie and ageless rock discernible over a long distance in clear weather. Here Strathfarrar is at its widest, a full ten miles as the crow flies from the serrated north ridges to the south marches and the great

reposing bulk of An Riabhachan, highest hill in Ross-shire. A vast
scene of high shapely hills entwined with river and lochan, woods and
corries which long hold the gaze of those with a feeling for wild places.
By and by one would look down to the more immediate surroundings
and the pleasant pine and larch bordered croftland of Monar itself. The
grey stone lodge nestling snugly amongst the trees; a burn, noisy with
falls and pools, at its side door. A thoughtfully planted avenue of pine
trunks framed the view from the front lodge windows. Out it led,
over to an old thatched boathouse at the margin of the loch, out and
away, away to the far peaks of stirring memory for the irresistible
stalking days.

Here then was the country in which we found ourselves, harsh
and kindly by turn, a land of deer and sheep, wild life in profusion; no
road nor telephone, only the boats, the ponies or two sound feet.
Perhaps our faraway home's most remarkable feature which we came
to value above all others was the quality of its silence.

SHEPHERDING WAYS

The Strathmore ground was noted in the early stalking books as amongst the finest hill grazings in Scotland. 'Deer Stalking in the Highlands of Scotland' by Lieut. General Henry Hope Crealock, certainly one of the most beautiful books I have seen produced on this subject, lauds the quality, grandeur and beauty of West Monar as the author saw it over a hundred years ago. A well known geological feature divides the Mica Schists of the east coast from the Lewisian Gneiss of the west, and passes north-south through the Strathmore hills. This belt of dividing rocks being of a calcareous or lime bearing nature accounts for the lush greenness of these pastures and provides for the wide variety of animal and plant life.

In the 1870's the ground managed to support a surprising 3,000 sheep, together with many cattle, perhaps about 50 head, as well as becoming a deer forest of considerable reputation. The sheep at this period were of a small unimproved Cheviot type. It was not the practice to sell lambs off the hill as we do today but rather to keep the wedder lambs and winter them on the low ground in eastern farming areas. In spring the wedders returned to the glen where they would become a separate hirsel. Here they were kept until three, four or sometimes five years of age before they were drafted fat to the markets of the south. The quality and taste of the mutton produced under this system cannot be found today. It has a flavour superior to any meat I have eaten. The wedder stocks also produced a heavy and valuable wool clip, mutton and clothes for the growing industrialised millions were the staple commodities of the hills and glens. Strathfarrar played its part and Strathmore followed the sheep rearing practices of the day for a span of a 120 years.

Strathmore in my day ran two hirsels of sheep. The north side of the glen held a flock of three hundred South Country Cheviots whilst across the river on the south side and along the extremely steep face of Meall Mhor carried a heft of 200 Cheviot cross Blackface. The latter were ewes originally of Cheviot stock which had been mated with a tup of the Blackface breed. Even under the less favourable ground conditions of a north facing hill, these crosses, possibly due to their hybrid vigour, fared better. They were more prolific and milked a better lamb. Further, it is fair to say that during a period of some thirty years, from the fifties onwards, the South Country Cheviot breed found itself in declining popularity. The market demand for a large carcase mitigated against these small hardy sheep. In addition flockmasters faced with their inherently lower reproductive rate found increasing shepherds wages and other costs could not be easily covered from a falling gross return. They are an extremely hardy breed with a low cost of production and as I write the small carcase they yield is once more in favourable demand from new middle-east openings in the sheep trade.

The five hundred breeding ewes over which I held charge involved us in the gathering of two major hill areas. The Strathmore north side gathering involved around twelve to fifteen miles of walking over high tops just short of four thousand feet. Strathmore cottage itself was a little above five hundred feet and the climbing undertaken put a lift in the step once you became hardened to it. To the south side of the main Strathmore glen the gather, being mostly on the one lengthy hill side of Meall Mhor seemed more straightforward and gave us less walking. It was nevertheless difficult and dangerous due to the steep and slippery nature of much of the ground. On this gather Iain MacKay was top man and followed the ridge and peaks using Roy, his old Skye collie. A large black and tan animal with a long head and powerful muscle, this rangy dog was a remnant of a useful collie type now almost vanished. Moving on a hill came naturally to him and Roy's loping, seemingly unhurried stride actually covered the ground deceptively fast. He didn't lie or creep about eyeing the sheep in the fashion of the modern border collies of T.V. fame. Walking up steadily to his sheep he would stand when required to stop, only lying down with reluctance and obvious disgust.

Roy was truly a dog of the high tops and the wide gatherings, and Iain clearing the ridges on a south side gather would make his veteran dog stand and voice ringing barks down the corries to hunt any top sheep away down to me as I worked the middle ground. This was a singularly difficult route, leading me along narrow ledges which, sweet with short natural fescue grasses, would tempt sheep into potentially dangerous positions. The shepherd required both a head for heights and a sure foot which as well as confidence in each movement were essential safety factors. Gathering such ground the dogs had to be sternly checked from advancing rapidly towards sheep venturing precariously for a selected bite. Instead much noise or stone throwing ploys proved safer, tactics designed to flush the sheep rather than drive them by force. On odd occasions too much force without warning caused us to witness a sickening spectacle. The unfortunate sheep in sudden terror at spotting a dog dived unheedingly off the ledge spinning over and over, falling perhaps two hundred feet to bounce and thud, down to where the ground sloped out. A dramatic lesson for the shepherd warning him to step carefully and avoid such a result. To this day I still can hear in my mind the sickening wool-muffled thuds of a fatally falling sheep. Sometimes, particularly Cheviots, if startled by a dog, galloped wildly down a steepish hillside. Whilst their front legs managed to hold a stride they lost control of the hind legs which hideously stretched out behind like the handles of a wheelbarrow. The crazed sheep would not stop and might reach the bottom in this condition, their backs broken. Recovery was impossible and out of kindness on the rare occasion it happened, I killed with a blow to the head from a handy stone.

Happily gruesome incidents of this nature seldom occurred and a smoothly moving gather on the south side of Strathmore filled a day with many pleasures. It combined the satisfaction of using one's skill in dog handling with personal strength and sureness of movement. Sandy the shepherd from the home estate refused other than the lowest routes but using his good dogs rattled the gathering bulk along the glen bottom as we turned each lot down to him. Though the best of shepherds he admitted to having no head for heights or airy ridgewalks. We would tempt him to ascend our Olympian heights with accounts of panoramic views from the tops. "...... you lot" he would say, "I'd rather the view from the bottom".

One south side gathering I remember, Iain resorted to his usual trick (a practice frowned upon by the hillwalking and climbing fraternity I might add) of rolling large boulders from the ridge in an attempt to frighten sheep holding refuge from our efforts on an impregnable ledge. Sandy, I could spot, two thousand feet below me, a minute figure strolling along on the easy lower ground, I dare say whistling a carefree tune, for it happened to be a sunny heart lifting day. At a shout from Iain I guessed imminent danger. Only seconds elapsed, my forewarning materialised as a large spinning boulder, crashed past just a hundred yards ahead of me. "Sho ho below" I bellowed to Sandy. The seconds it took for my call to reach the blythe shepherd proved almost fatal, for the whining missile hurtled towards him. He stopped. I saw him look up. Instantly he dived for the safety of a large stone which by good fortune lay at hand. From my now complacent vantage poor Sandy looked for all the world like a tiny terrified spider scuttling for its corner. The boulder crashed past his shelter stone creating a mighty splash as it plummeted into the river. Minutes passed. I focused my glass on his refuge. A face peered around the side, followed by Sandy who judging the air raid to have passed, emerged from hiding. He stood waving and screaming up at us. Words and dubious imprecations floated upwards in the quivering convection currents of that sun blessed day. The exact words of that shepherd's displeasure, I will leave to your imagination.

Talking about shepherd's use of bad language reminds me of an occasion when young Kenny MacKay happened to be out rowing for a most impeccable Gentleman who quietly cast a fly on tranquil waters out from Strathmore pier one unblemished summer's day. A shepherd happened to be working sheep on the hill face above our cottage. All was not going well between man and dog. He opened the commentary by referring, at the top of his voice to his dog being a "black bitch of hell". Developing the theme in a mixture of Gaelic and English he warmed to a description of the dog's character and apparent shortcomings. Voices carry far in silent country. The acrimonious conversation was clearly heard at the boat. Kenny, not knowing which way to look, told us the Gentleman began to blush and finally turning with acute embarrassment observed, "Well Kenny, I once said 'damn you' to my dog but I've never before heard language like that in all my life". Kenny laughingly remarked to us in recounting the story "Isn't it lucky he didn't understand the Gaelic".

The critical point at any hill gather focused invariably on the tactics adopted by the shepherds on reaching the shoulder of some of the large hills. If the hurrying sheep managed to gain a curving shoulder unchecked they made smartly round it and then away like the mischief into extensive hidden ground always safely out of sight. Hours later the truants could be spotted sneaking back to their own bit of hill when the gather was well past. To prevent such a disaster, Iain MacKay from the top of Meall Mhor on a south side gather, would gauge the progress of men and oncoming sheep. At a crucial moment he sped down the ridge, stick waving, glass bumping on his back to hold the sheep on the east end of Meall Mhor in a scooped out basin known as Corrie Dun Beac. Here the thwarted sheep held by a panting MacKay, gathered together milling and bleating as if discussing where the next escape route lay. Closing on them Iain and I commenced the drive down to the lower slopes where Sandy would have the bottom sheep rambling along calling for lambs and generally protesting. The process seen from a distance must have resembled lots of little snowballs rolling together as trickling lines of sheep starting from the west, grew together in number, until we joined them into one white, undulating mass on the eastern ridges.

Having pushed the bulk of the sheep together and down to the lower slopes of Meall Mhor we forded them across the Strathmore river in order to reach the path for the long drive down to the fanks. This forced swim often necessitated some force by dog and shepherd. A small cut, literally caught and thrown into the current made a dive for the other side, and the remainder grudgingly followed.

Already wet or hot we thought little of plunging across after the sheep and in all the years of wading or wettings I didn't suspect it to be the cause of any cold or complaint, nevertheless it was obviously best to keep on the move. The bulked gathering we pushed down the pony path to the home fanks at Strathmore. This final stage on a hot day found the odd sheep tiring. Some in poor condition decided to flop down with head and neck stretched out, refusing to move further. All manner of tricks were tried to goad them along and if all else failed we tipped a handful of water into their ears. Occasionally the torture succeeded but once a sheep lay down it really intended to become a problem. Apparently breathing their last they prostrated themselves at the side of the path. Left for dead, a glance back after a hundred yards often revealed the corpse bolting back for the heights. Fortunately this cunning practice did not spread through the flock, which must indicate the limitations of their vocabulary.

We forced along the active sheep, and from a considerable distance their bleating progress was loud enough to warn Betty of the approach of men ready for a hearty meal. Lambs darted this way and that having lost their mothers in the birly tangle, the dogs dashed from side to side to turn in breakaways and press the white undulating column homewards. Reaching down to the west gate of the Strathmore croft brought the last critical test of a gather. At this stage many lambs separated from their mothers fell to the tail of the procession. Instinct strong in their behaviour, told the agitated creatures that mother had last been bustled away high up Bidean a Choire Sheasgaich or wherever the ewes' territorial haunt might have been. Particularly on a day when blinks of sun cast changing light on our labouring scene, the group of flagging lambs, sometimes thirty or forty, having refused to follow the ewes through the gateway found new energy, leaping over dogs, catching sticks and swearing men in their race back up the path to freedom. Pursuit of fleeing lambs at the end of a hard day became the final exhaustion. Distraught ewes began running after their disappearing offspring. It was assuredly a critical point in the day and it took much skill to steady the situation and not lose all.

Finally, with ewes and lambs secured in the pens and limping dogs stretched tired at the door we trooped into the stone floored kitchen ravenous for the awaiting meal. In highland custom our table accommodated anybody who came by at a meal time. Soup and venison stew could never taste better than after a long day sustained only by a cheese and oatcake piece. Whilst weather-faced gatherers ate hungrily, discussing the incidents of the day, the children toddled up to the kennels with a heavy pail between them to feed the dogs. Sandy sometimes stayed with us but the MacKay boys after a good sit down would rise stiffly and row home to Pait. If the night kept fine we would arrange to go down to the loch later that evening for an hour or two at the trout fishing. For us only daylight counted, time ran the outside world. We'd pull leisurely across to the mouth of the Strathmore river and if the fish were taking then a happy night's casting and yarning was ours until we could no longer see the flies.

★ ★ ★

A GOOD collie with confidence in its master will work away from him at great distances on the hill. An almost telepathic relationship would seem to function between man and dog on some occasions. The Strathmore hills, steep and high, lent themselves well, indeed often required using a dog at over a mile distant. Sometimes the dog could only be followed with the glass and the need for a black dog with a broad white collar was obvious to me.

Nancy who became the main dog I used for ten years was beautifully marked. Her broad white collar and black curly coat always carried a shine. The tip of her tail, legs and underside were white, together with a blaze up her muzzle. Ears that normally lay could be cocked in interest but the tips always turned down. Nancy as well as having good looks was also of a great nature. I never once saw her snap no matter how the children teased and tormented. Her character and intelligence came through in fine dark eyes. In Nancy's best years distance was no object and she would work away gently at complicated handlings without my instructions. Until she had built up such confidence however, there were a number of instances when she seemed to be able to read my mind.

Putting her out one day on the south side of Strathmore to climb the very steep slopes of Meall Mhor, I watched her move steadily, making a wide arc around a cut of unclipped sheep I wanted brought down. Taking an eye off her for a second she slipped out of my sight. The sheep didn't move and no amount of shouting commands or coaxing words had any effect on the dog whom I guessed to be lying. I sat down and spied the likely area with the glass. Luckily after some minutes I spotted her black and white face peeping over a rock as she intently watched the sheep which remained undisturbed, grazing some considerable distance from her. Upon sighting her I immediately called a command. Straight away she started to move again, coming in on the sheep and wearing them steadily to do a first class job of taking the cut down to the flats. At a distance you can generally detect the dog's whereabouts by the behaviour of moving sheep. Nancy seemed to sense, not once but on many occasions, when I lost her to view. She would stop until I spotted her. The moment I did so all was well and off she would go again. Quite often the instructions would remain the same, I always tried to whistle or call as though I were seeing her. Invariably she seemed to know whether or not I had her in view and no amount of trick calls would have any effect. Nancy came as a puppy from a famous border breeder and I found that the most successful dogs were invariably the ones I reared and trained myself. Complete confidence came to this dog by the time she was two and a half and a good example of her abilities happened one fine spring day.

It was early in April that I got word the ewe hoggs were coming back to the glen from their wintering quarters down on the home estate, or 'East the country' as they say. Taking the *Spray* down to Monar I had Nancy with me and arrived to find the lorries about to unload some 150 hoggs. Either Nancy and I must walk the sheep home, some seven miles, and then return for the boat or I must put the dog to the test. The pathway home was a rough pony track, which for its first couple of miles was in pine woods and well away from the lochside. Deciding to try the dog I set the sheep off through the iron deer fence gate and away from Monar pier. Nancy got firm commands for driving sheep before her "walk up, walk up, good dog walk up". Hoping she would understand the plan I stood near the gate way and encouraged her to force the sheep away along the path. Several times she looked back as the distance separating us increased,

but kept on working the sheep away on the westward track. Before she could look back again I ran to the boat and set off through the 'narrows' at the foot of the loch en route for home. Curious and perhaps a little anxious I landed at a point on the lochside where the path eventually came close to the waters' edge. Mooring the *Spray* to a rock I went ashore to see what might be happening. I hid behind a tree so as to see what she really did without me. Some little time passed and then to my relief along came the leading hoggs moving quite smartly. From my hide I saw that the dog was in fact doing her job splendidly and padding along quietly behind them, keeping the flock moving steadily, but without the hurry that would have split them up. She was keeping them on the track and moving the side stragglers as necessary, all in fine style. Attentive to her task she was quite close before spotting me as I stepped out from the trees. The relief could be read on her face. I got a dancing welcome. She had made the grade. I set her off again, went back to the boat and up the loch to a point near home where I landed to open a gate into the hill park. Some two hours later Nancy and the hoggs arrived home. From that day she went on to do many tasks which required her to anticipate and act on her own initiative. Seldom did she do anything stupid.

Sheila was a much older dog and had been through one or two hands before coming to me. A bitch of great strength, perhaps rather short of leg, but with a wide chest giving plenty of heart room. She was perhaps the best dog to stand heavy work that I ever owned. With sound hard feet Sheila would often gather for three or four days at a time with a spell of fank work to follow. The areas we covered both on our own hirsels and when helping neighbours were unquestionably amongst some of the highest, steepest and most extensive in the Highlands. As Nancy improved in her capabilities and got more to do, so Sheila became jealous. One day during some routine work west of Strathmore whe went off on her own and did not come back down the glen with me that evening. Darkness fell, still no Sheila. Later that night we were sitting reading in the kitchen when sharp barking caught my ear. Without waiting to put on a coat I hurried out. Sure enough going over to the deer fence which surrounded our croft, I could make out Sheila barking wildly at a small cut of my own sheep. A call did not stop her. I went up to find that she had killed three ewes up against the fence. She was tearing frenzedly at a fourth. Only with difficulty could I stop her. I instantly destroyed the mauled sheep. Sadly taking Sheila down to the house, I found a chain with which I tied her to a post near the kennel. Taking the rifle I shot her there and then.

Both these bitches had exceptional eyesight. They could spot sheep at up to a mile as was often obvious to me by their actions. Neither of them however could match rascally Bob the Beardie. Bob came from Argyllshire and, I was to discover, brought some highly indifferent habits with him. These shortcomings were frequently brought to his attention but in character he could be likened to a truant playing schoolboy. He hung his head, promised not to do it again, but couldn't resist the temptation when bird nesting times came round again. Never did I have a dog who could better portray shocked innocence. When getting a lurid cursing he did not cringe but would sit bolt upright, tufts of his long hair hanging over the widest unblinking eyes. Slowly he'd turn his head to gaze away to the distant hills – "who me?" "Shepherd would I do such a thing?"

Bob, a black, rough coated dog was of a type known as a

'beardie'. Such a dog barks to order, drives sheep ahead of it and is excellent for handling bulk gatherings. I soon discovered that Bob did not relish being kept in a kennel and in a single night could easily gnaw his way out of any wooden structure. Chaining had the effect of producing a night's barking. After being a week home I gave up. Bob was left outside and took on the job of guarding the home, making his headquarters under the paraffin shed at the end of the house. One of this ruffian's main enterprises in life was to achieve the status of combatant par excellence. His eyesight was exceptional and out on general duties he would while away the day gazing across the loch to Pait - something over a mile distant as the crow flies and clearly visible. Any enemy movement he detected would trigger a barking session from the sentry box at the garden gate. He would then watch Pait pier and should the movement suggest somebody getting into a rowing boat he would tear down to our pier to stand guard, barking furiously as they pulled towards him. The MacKay dogs were reluctant to disembark until Bob had been driven off with sticks and stones. Like a seasoned combatant he was singularly adept at dodging a swinging boot or a well aimed stone. Once the Pait intruders landed a dog fight would always ensue and sometimes two or three before we could get off to the hill.

Out and about his work he was the hardiest and most useful of dogs. He would climb up the roughest face, stop and look down to me and at the call "Speak up Bob" he would give a salvo of barking which would set any gathering scampering for their lives. Working in behind sheep he would trot along for any distance and could then be persuaded to half turn them down the hill. Get the bulk of sheep together and you were home in half the time as he tirelessly 'woofed' them along the path.

To my cost he discovered that the deer ran faster and further than the sheep. They were much more challenging and entertaining. If he encountered deer whilst being directed towards sheep then no amount of shouting or cursing would recall his attentions to the path of duty. He considered the gathering must proceed without his services, a day of kingly sport was at hand.

This waywardness could largely be indulged but then he finally blotted his copy book. The MacKays and myself were stalking hinds for essential winter salting out on the slopes of Sgur na Conbhaire. Positioning ourselves with great care after a tricky stalk we moved in on a large herd of quietly grazing deer. Our rage knew no bounds as the hinds suddenly were stampeded by a black object trailing a large length of chain, crossing the skyline. Indeed such was my chagrin I fired a magazine wildly at the villain. My rage, the distance, and his charmed life kept Bob safe once again. Although left chained at home, he had decided the day could not possibly succeed without his assistance.

Matters came to a head however when one day, in his usual cocky style, he set about Kenny's poor old Spot whilst we were working at the Strathmore fank. A serious fight ensued which was only broken up by much beating and shouting. Bob was the transgressor and therefore the chief recipient of our blows. Finally he fled, with me in pursuit, to his den under the paraffin shed. I reached into the bolt hole and dragged him out by the scruff of the neck. A gentle old maiden aunt who happened to be staying with us appeared at the front door to see what was causing the stir. Now our task that day had been lamb castration which was effected by the old method of knife and teeth which smeared my bearded face with blood. Still shouting, I proceeded to hold him out at arms length and wollop him with my stick, also aiming a kick by way of good measure. Quick as a flash, with great dexterity, as befits any good fighting man, he twisted round and bit me through the wrist. My blood stained face contorted with rage and pain, I threw him spinning into the air. The dear Aunt blanching at this primitive scene, retired inside to sit down. Bob made off and lay away for several days no doubt spying on us from a distance.

That was enough, an advert in the paper was the only solution. An unsuspecting farmer from the Black Isle wrote saying this could be the very dog he needed. I took battler Bob on a chain to Inverness and over the Kessock Ferry. Walking him up the pier at North Kessock to where the farmer awaited us in a suitably large car we chanced upon a mongrel dog which for no obvious reason was loitering about the street. Before I could take evasive action a ferocious fight broke out. I was in a dangerous position holding the chained fury. The Farmer in innocence and not knowing the character involved, attempted to separate the combatants with his bare hands. Fortunately the cur was no match and fled. I handed Bob over to a new owner with the passing caution that though he was an excellent worker he had 'something of a fighting temperament'. I didn't hear more of dog or farmer but was to be left no uncertain legacy by the invincible Bob.

★   ★   ★

THE morning sun lifted the early mists that spangled the hillsides and swirled out through the high corries of An Riabhachan. The hill's long spine lay in my view away above Pait as I looked out from the byre door whilst milking the house cow. Alison was feeding the hens from a wee pail. I heard her at the end of the shed imitating her mother and calling "tuck, tuck, henny, hennies". Hector sat on the step of the byre poking amongst the stones with a stick. I watched that nothing went into his mouth. He wasn't long walking and most objects required tasting. The lifting mists were a sign of fine weather; we should be gathering today I thought, and the boys will shortly row across. Walking back to the house with the milk pail, sure enough I saw the two MacKay boys and Sandy, the home estate shepherd, coming down to the pier from the little cluster of buildings across at Pait. Half a mile of track led from their house down to the boat sheds. Shortly after I noticed the rowing boat pulling smartly towards the Strathmore side. I knew from the short hard strokes that Iain would be at the oars. Five minutes later they came striding up to the house, sticks and telescopes their equipment for the day, dogs running excitedly about their heels. There was the expected ritual sparring with my two growling dogs as the boys came to the door. "A grand day for the hill" I greeted them. "Yes indeed I think it should keep up" replied Kenny. "If it holds till we get a cast on the loch tonight it will do", Sandy seemed more enthusiastic about fishing than gathering. We went on to discuss plans for gathering the north Strathmore cheviot hirsel.

Our deliberations over coffee were cut short by yelps and snarls. Sandy's Moss and Kenny's Spot snapped at each others' throats. The bitches looked flattered. It was time we moved for the dogs were keen to be off. Being my ground that we gathered, and one of the two hirsels which I shepherded, it was expected that I do the most climbing. A good pair of sprung hill shoes and rain-shedding tweed plus-four trousers, glass on my back, and a venison piece in my pocket, I took stick and dogs to the long steep ridge above our house.

The Bidean an Eoin Deirg top was my first objective. I spanged it out with a long unpausing step. Some twenty cheviot ewes and lambs were however, scattered on the far side of the deep and wide Corrie Shaille which lay below me as I climbed. They grazed on the rich green slopes below the scree skirted shoulder of the Bidean. I sat and spied, steadying the glass against my long shepherds' crook so as to get a clear look at the ground. An essential precaution before putting off a dog to move the sheep and so start the gathering in motion. The wind blew cool in spite of the mid August sun. The air was clear and the view sharp. I noted the line a dog must take to clear the corrie, shut the glass with a click and stood up. The dogs were instantly alert, dancing about me with excited anticipation as they awaited a command. "Nancy sit down, come in here, sit down I tell you". She reluctantly complied. "Sheila, good dog, go way off" I gave her a wide cast with a wave of my stick. She ran a hundred yards and then stopped to look back. "Good dog, way wide". Sheila knew from the words and tone that she was on course. Realising the work expected of her and knowing the terrain, she set out deliberately for a long hard run. Nancy, the younger dog I kept tight to my heel, avoiding running two dogs out together when on the open hill unless both could be easily seen. Sheila disappeared over the ridge away to my right. With encouraging shouts at intervals of "good dog, keep wide" I waited and watched.

The sound of the tumbling burn in the corrie bottom came and went as the wind swung round, strengthened a little, then died. Taking the glass out again I spotted a group of hinds grazing quietly amongst some rocks a little distance above the scatter of sheep. As I studied the deer, noting a number of strong-looking calves I saw one of the hinds stare hard at the ground behind her. Swinging the glass towards where I thought the hind's gaze indicated, the dog came almost at once into the telescopic field. Good old Sheila, she had made the far corrie face. Putting down the glass I caught sight of her with the naked eye. She would be about three quarters of a mile from me, straight across the deep intervening ground. Glancing back to the hinds, I saw their heads jerk up, they paused for a second or two and then began to climb rapidly towards the shoulder of the corrie. My attention now focussed between the dog and the sheep. "Steady Sheila, steady, steady" I called. It was always interesting to note the time lag over a distance between one's word of command and the dog's reaction. Seconds later Sheila paused. The wily ewes having spotted the deer moving out now searched about them to find the cause of alarm. The combination of hearing my voice and sighting a movement of the dog some two hundred yards to their left, set them running smartly off across the corrie face. Their scheme I knew would be to outpace the dog to the lip of the basin and then disappear effectively amongst the steep ground to the west. "Sheila, way off, way off". My tone carried urgency to the dog and in spite of an arduous run she picked up speed and cut out above the hurrying sheep. Positioning herself ahead and slightly above with the experience of many a hard run Sheila checked their flight. I let the sheep and dog steady up for a few moments before changing their direction. "Come in ahint them", I directed here. She came in behind them and turned the group neatly down the green, stone dotted face of the corrie. The sheep were moving too hard for their safety. I checked the dog with a single sharp whistle, two little fingers into the mouth and a blow as hard as possible. The dog lay down.

Knowing Sheila now to be in command above them, the ewes kept on down the Corrie Shaille face to the lower slopes of Toll a Choin, the main valley running away to the Bealach. Iain MacKay following round below me in the middle ground would find these sheep before him and join them to the group he moved westward up the valley. Two sharp single-pitch whistles gave the signal to Sheila to come to heel. I watched her set off towards me and knowing she would reach me in due course recommenced climbing the ridge. Sheila had no small climb to regain my heel, nor would that be her last run for the day. I neared the high top before Sheila caught up, tongue out and panting for she was a heavy coated dog.

Fine and fresh was the day filling the lungs with pure mountain air. Lightsome came the step with an exhilaration of bodily fitness. The great bond between man and his natural environment felt never more real. Looking down to the tiny lochan below I took pleasure from its colours. Set in steep green walls and having a bottom of white mica sand the effect of the sun's rays slanting over my ridge and plunging shaft-like into its secret depths was to create a turquoise jewel. The winter reflection in the wee lochan set all a cold sparkle.

"Spot, get away back up. Get awa-a-ay up Spot". Young Kenny MacKay's voice came drifting up to me from a thousand feet below. I now stood on the ridge of Bidean an Eoin Deirg, shepherd's stick bracing me against a strong, snatching wind that surged over the precipitous rock face which fell away at my feet. My narrowed eyes searched the ground far below for a movement. Quickly I caught sight of Kenny's swinging step as he crossed a burn flowing out to the lochan in the deep crater hollow to the north side of Bidean. He stopped and I watched him give a directing step to the dog together with a sharp wave of his stick. Seconds later I caught his voice again, "Spot you fool get back out of that". I knew his dog must be working somewhere on the difficult rock face between us. My two bitches Nancy and Sheila looked down, bright eyed and excited, then back to me for a word of command. "Come here and settle yourselves" I warned them. Momentarily subdued by the tone of my voice they sat, but not without an apprehensive glance to check if I were really annoyed, for it often seemed that the dogs could read the look in my eyes.

Myself, and my two invaluable team mates, looked carefully over the sheer edge of the ridge. Kenny had stopped walking and now an angry shout accompanied his waves. As close as I dared go to the thousand foot drop, I searched for his problem. Moving twenty yards up the ridge I spotted a group of six ewes. Two of them still unclipped, stood on a rocky ledge defying poor old Spot who not having run sufficiently high and wide had come up to them from below. He moved a step or two at Kenny's by now enraged urging. The leading ewe snorted down her nose and stamped a front foot at him. Spot lay down. From bitter experience I know that at this point the shepherd also feels like lying down and perhaps beating the ground with frustrated rage.

I moved into action and climbing another three hundred yards up the ridge came to a point I guessed roughly to be above the recalcitrant brutes. "Shaho, shaho" I bawled at the top of my voice, at the same time selecting a large boulder at the very edge of the dangerous face. A good heave with the foot, and taking care not to follow, I rolled the large stone off down the precipice towards the defiant sheep. They had now seen and heard me. The spinning projectile crashed down behind them bouncing from ledge to ledge, sometimes clearing two or three hundred feet at a time to burst shell like perhaps a little closer to young Kenny than a safety officer might approve.

The troublesome group thought this treatment enough and moved smartly from their previously invulnerable ledge. Spot unperturbed by the falling missile now moved cleverly in behind them and without coming too close commenced a steady drive which brought the stubborn sheep clear of this treacherous area.

I climbed out over the high peak knowing that both the boys worked away to either side of me on this great thrust of aged landscape. Iain down to the south on the grassy middle slopes of Toll a Chaorachain moved straggles of sheep with his noisy springy little dog, Bess. I could see the lines coming together at the head of this long wide corrie, taking tracks, time honoured by generations of sheep and deer. The beaten path avoided climbing the steep snap up to the Bealach itself, instead cutting away to the right in a more gentle incline out through the Bowmans pass and over to the greens of Pollach an Gorm.

Sandy far behind, with his dogs Moss and Coon, now headed away for the west shoulder of Sgurr na Conbhaire, there to lie in the sun waiting, ready we hoped to hold and turn the bulk of sheep we would eventually drive towards him round the curving face of that hill.

Kenny working to the north had the longest walk and had been out of sight for quite some time. I knew the area he covered, though it held few sheep, was extensive far flung ground, and would involve him in two or three extra miles walking. Iain and I met on the Bealach. The dogs, after wags of recognition lay down, legs stretched, flat on their sides. Against the strong sunlight I could see the single file of sheep winding out to our left over the spectacular Bowmans pass - a high and narrow fault formation which slashed across the ridge like a giant hatchet blow and had obviously been used by cloven feet for centuries. The midday sun felt warm on the rocks where we sheltered from the remarkably penetrating wind. We picked a sun filled corner out of the wind and lay down, faces up to worship its comforting warmth. The day drifted away to the clouds on the horizon and we waited. A rattle of stones and Kenny's voice, "You lazy so and so's" brought the day racing back. A lean shepherd stood looking down from the edge of our suntrap. Jacket off, sleeves up, face red from the long pull up the north side. "Ah well boy, isn't this just the life" I mocked opening an eye but not moving a limb, "great the pleasure of simple shepherding duties". We watched the cut of sheep he had gathered from the north side of our ground, file across the Bealach a little below us and make away to join their flock mates trekking through the Bowmans pass. Reluctantly we rose. More climbing and running lay ahead, before the gathered lines of sheep swinging round the Sgurr na Conbhaire face would, we trusted, waken Sandy.

Incidents throughout a hill gathering involved an interplay between the shepherd and his own dogs, together with team work amongst the men concerned in gathering the sheep stocks from these vast and often desolate areas of the Highlands.

Much could go wrong as you could well imagine and I use the word 'wrong' without exaggeration. The MacKay boys were widely known for their running abilities and over many years had won the hill races at various Highland games. I had no option but to emulate their ability to move swiftly about the terrain. During the summer we generally wore plimsolls at the gatherings, for their lightness added fleetness. Often we had to run if the dogs got tired to a degree where they refused to run out for sheep or just couldn't find the energy to climb again. The gaunt empty-sided brutes would run a few yards then flop down. More aggravating for the tired shepherd, some hard pressed dogs would sneak away for a rest and remain wisely out of sight until the day passed. Looking back I believe that our ability to run and climb fast was a factor that made sheep farming possible in this difficult area.

One long day, leaving early to gather with the crofters from further west at Achantee, we walked or ran a good thirty five miles, within which distance there was much climbing and descending. The dogs became lame and stiff on such a day, their feet inflamed and scalded, red between the pads, sometimes bleeding. They flopped down exhausted, legs stretched out in each burn or bog, lapping at the water. We also were glad of a sit down, though a drier one if possible. A good hillman normally has a very long stride. He picks his route almost by instinct and doesn't stop unless to direct dogs or spy. His step will spring off the toe into an easy stride. Climbing a steep pull he keeps close to the ground and with flexed knee appears to float without apparent effort, sometimes to the surprise of well equipped and trained 'mountaineers'. Going down hill, except if dangerously steep, saw us at a run. This came easier on the back of the legs. "Not so hard on the elastic" remarked that hardiest of men, old Duncan Fraser.

Whatever might be the trials and tribulations, the shortcomings of shepherds or dogs, clear weather was essential for a clean gathering in such high country and many days tended to be lost waiting for mist or rain to clear. Given a good day one of the special pleasures of gathering on Strathmore were the sweeping views presented. From Sgurr a' Chaorachain on the curving north ridge an impressive scene reached away westwards. Nor was there lack of beauty in miniature, it was scattered profusely under one's feet. Rare and charming on the Strathmore tops was the Alpine Saxifrage. On many of the heights and ridges the rock formations threw up vertical strata. Veins and fissures would be weathered by frost and rain, almost to a shingle of shattered stone. Amongst the thin moss and lichen turf overlaying much of these areas one found large patches of this hardy little Saxifrage. A charming gem of the Strathmore tops, hugging the ground for its very life, the tiny flower of bright crimson and white measures less than an eighth of an inch. They cluster in a thick mat surviving on some of the harshest environment found in this country.

Looking out over the near but lower hills of Achnashellach, range upon range unfolded to the horizon. Far above Loch Carron the gentle hills of Coulin estate lay dwarfed below the strangely light coloured Torridon massif. Beinn Eighe and Liathach often show white due to their sandstone rock formation. The colours of each range would alter in relation to the distance and strength of light. Black and ultra marine through to the softest of fading powder blue. Coming round a point on a truly clear day, perhaps in May before the rain, far out across the Minch, the eye could just catch the pencilled outline of the hills of Harris, a distance approaching one hundred miles. The particularly fine Applecross range were often in view, framing a prospect of the Isle of Skye, which lay in pristine beauty directly away to the mysterious west.

The low lying hills sheltering Plockton Bay led the eye across shimmering waters in the Sound of Skye, over the Islet of Pabbay, to the majesty of sea skirted mountains. I would never fail to search the Sound for the sighting of a ship. One day I was memorably rewarded. The odd, almost spectral view of a fine three masted schooner under full sail came into my glass. Her white sheets moved gracefully against the dark shores and fingered headlands of this island of mystery, for on that day Skye looked almost black. The jagged peaks of the Cuillin gave an impression of gazing out to a Fairy castle, as a schooner sailed in from a bygone era.

Many a late evening in high summer twilight I would climb out to this, my favourite vantage point. The dark red sunset haloed behind the far Cuillin ridges, brought them to touch one's feet as the burning rays created an eerie silhouette of these ancient hills. The cold grey Atlantic waters were lost to the sky in the warmth and colour of the sun's parting light. I stepped quietly home in a world of silence.

REMOTENESS LIVING

"Cows Nancy, cows, cows, good dog, fetch the cows". Nancy, dancing about my feet with glee at this command would give little yelps, then looking intently west from the croft to the Strathmore river flats would spot her charges grazing on the natural but lush summer greens. Without further instructions, off she'd speed, running along the twisting river bank, excited eye on the cattle, not looking back to me. For a house cow we had Margaret, a brown and white Ayrshire. Her companion, a cross cow which we didn't milk was left to suckle her calf. With no measure of originality she was called Highland Mary and both were purchased from Skye. In the summer days after morning milking they would be driven out at the west deer fence gate, and at their leisure ambled down to the fertile river pastures, browsing natural grasses and herbs, with no artificial fertiliser to marr their sweetness.

Looking after the little herd and milking the cow was our chief daily discipline. I will admit we rose in the mornings according to the weather or our own inclinations. Occasionally some particular arrangement might be made, but even this was in our favour. Without a long walk, no one could reach us and were therefore unlikely to arrive inopportunely. The Monar family over six miles east by water were our nearest neighbours in that direction. The MacKays across the loch had similar inclinations, to our own, therefore our pattern of rising was unlikely to be disturbed. Twelve miles west of us out through the hills lay the next nearest habitation. Alistair and Flora MacRae at Iron Lodge, born and brought up at Strathmore when their father shepherded for the estate, certainly did not constitute a strong threat to our peaceful, not to say indolent morning contentment. All in all we felt relatively secure and from this happy position largely did as we pleased.

The cows perforce had to accept a flexible routine. So after supper each summer's evening Nancy would be set off to bring home the cattle. Because the flats were criss-crossed with ditches and drains she had sometimes to drive the cows away from the croft to a safe crossing point before heading them on the easting homeward path. Given enough time, a combination of dog intelligence and bovine instinct, the cows would arrive to await us at the west deer gate. Nancy dawdling down to the house door indicating her job was complete meant bestirring with milk pail, washing pail and udder cloth swung on either arm out to the byre. The evening's milking might be anytime from seven to midnight, as the long high summer days see little darkness in the Highlands. When Margaret, a spring calver, went dry towards February each year, we got milk from Pait and they in return would share with us in the autumn. The taste of milk produced from these untreated pastures had a sweet tanginess that I have never experienced before nor since.

Similarities in outlook and lifestyle down to small detail engendered good neighbourliness. Each day according to weather and season, daily tasks on either side of the loch generally coincided. Even the dogs were expected to conform to the pattern. Young Kenny had a grand old dog, Glen, which he had bought as a youngish animal from Fairburn, our home estate. Bringing the dog to Pait by road, car and boat, he didn't dream that when he let the dog go in the yard that evening that it might take off. Glen vanished surreptitiously. Due to the nature of communication it was some few days before Kenny learned that the dog had made off for its own home arriving by the next morning. In one night Glen must have either gone round or swam across the loch, travelled the thirty miles through rough hill country he had never before known and successfully sought out what he considered his rightful residence. Such navigating instinct and orientation must give rise to questions of presently inexplicable animal senses but of course Glen was indeed a wise brute.

On a fine fishing evening after a day on the loch Sir John and friends were strolling up the track from the pier to Pait Lodge. The

MacKay's house and steading lay in a little sheltered cluster below the lodge and from this now established home Glen came trotting down the path towards them. Sir John knowing the dog well, spoke. "Now Glen I'm sure you are not supposed to be going anywhere tonight. Come along home with us like a good chap". Glen normally friendly, merely went off the track and with a disdainful look resumed his progress down the path. The friends sauntered on up the track admiring the splendour of deepening evening colours over Ben Bheag, when Sir John with a touch of concern happened to glance back. Here were the MacKays' cattle coming steadily up the track from their daily grazings on the lochside driven by Glen. The group stood aside to let the cattle pass. Glen this time came across to them all tail wag and smiles. "I saw then" commented Sir John "that Glen was wiser than us".

Traditional to the Highland way of life, milking, cheese and butter making fell without question to the womenfolk. Mrs. MacKay invariably milked the cows, in addition to keeping the house, helping with the peats, the haymaking and numerous other outside duties. Old Kenny I never knew to do other than feed the cattle. The boys would sometimes take a turn at the milking when not busy at sheep or deer. Obviously the pattern and order of family work and duties were in direct descent from the deeply ingrained styles of the subsistence life implicit in the sheiling era.

The Kintail crofters sheilings a hundred years before had been out on the rolling spring fed pastures at the foot of An Riabhachan. We often came upon mounds and hollows of regular formation, certain traces of the humble turf built huts. Old Kenny following a wounded stag one day at the foot of the Green Corrie which is in the vicinity of these sites, noticed a disturbance where the stags had been boring in to the peat bank with their antlers. Here he spotted the blackened end of a small keg. This turned out to be a firkin about the size of a jam jar and which had contained butter. With a knife he prised off the lid, and the traces of butter seemed quite fresh. The butter made during the summer sheiling days, had been stored in the deep peat and as the sheilings were last used about 1830, it was a remarkable find. The transhumance of the sheiling system gave a poor but happy life as song and legend tell. By contrast our butter was made in a wee hand churn perhaps once a week. There was always as much as one cared to eat and for good measure I generally drank about two pints of milk a day. The crowdie cheese was a great favourite and along with the home made oatcake straight off the girdle, what more wholesome food could be wished, especially coming home hungry from the hill.

Getting the cows in calf so as to calve at suitable times of the year was of crucial importance to our self sufficiency system. We preferred early spring as the cows would milk cheaply and well over summer before drying off naturally sometime early in the New Year. One evening in late May I suspected that Highland Mary was in season and thinking that even in the hills we should move with the times I decided to take her down to Monar and summon the Artificial Insemination. Guessing she might be difficult to drive I felt a halter would be the surest method of coaxing her along the seven miles of pony track down to Monar. It was immediately apparent she had not been haltered before but with my wife's help we got her past Strathmore Lodge, through the east gate and out onto the track to her appointment. With unladylike if not downright indecent haste she took off without warning at a full gallop. I hung on, knowing that if I lost the rope all was wasted effort as she might circle for home. It so happened that the loch contrived to be in high flood that night and the path round various bays was well under water. Without slackening her speed the wretched beast plunged across the bays following the submerged path and literally towing me through the water each time I lost my footing. The river at Luib an Inbhir, half way to Monar was swollen, but still in what I presumed to be a state of thinly disguised anticipation, this she crossed in bounds and leaps. Here at last for safety I let the rope go, crossing more circumspectly than she had done. Suffice it to say I got the panting nubile creature to Monar and shut her in a park at the byre. It was now past midnight and of course there was no telephone. However my plan was quite simple. I walked the three miles down to Braulin, hoping and guessing that John Venters, the Lovat shepherd, would be going down the glen next morning. I left a soggy note on his door asking would he please phone the A.I. and direct them to Monar?

Walking back to Monar it occurred to me that on discovering the location of Monar the A.I. services might in future prefer me to make other arrangements. However I left a second note to Allan the keeper, explaining the case and asking him to kindly put Highland Mary back on the pathway home when all was complete. Then at last, it was home by the path to Strathmore.

I remember to this day how I enjoyed the freshness of that early morning. Every hillock seemed to be the territory of a Meadow Pipit. Their song and little rising display flight, though not as stylish as that of their relation the skylark, seemed fitting enough to the scene. The noisier Stonechat also took my attention, and with good reason, for his alarm call easily imitated by striking two small stones together came sharp and strident through the crisp air. A pair of Stonechats were obviously planning to nest in the stones behind the ruins of what we called the 'Halfway House' at the mouth of the river which I had gingerly crossed some hours previously. The male, with his black head, white collar and russet chest, a bold little chap indeed as he scolded me from a pile of rocks that shivering, light-intense morning.

Highland Mary wandered home latish that evening calling at the Lodge gate to announce her arrival. Needless to say my twenty mile trip was in vain, as she came into season three weeks later. This time I decided however she must contain herself and wait for the estate cattle coming west with a bull during June.

Fish formed an important item in our largely subsistance diet and by various ways they became welcome fare. The little loch below our home was a haunt of large pike. In the old days the Proprieters would employ fishermen especially to catch and hopefully rid the loch of this terror of trout and duckling. The last of these men at Monar in the 1930's was a Mr Prior who, something of a gentleman, stayed in Strathmore Lodge with maids and a ghillie passing his days trawling, fishing and netting these voracious fish. Myself and the Mackays requiring neither maids nor ghillies merely caught them for sport and a change of diet for the house.

The little loch lay under the shadow of the bulging shoulder of Meall Mhor and close to the Strathmore river mouth. Here were wide beds of reed and sedge, which together with overhanging river banks gave the shaded shelter in which the cannibal pike invariably lurked. Taking an old mesh net, on a selected evening, when the loch was at a suitable level, we'd row quietly to the channel mouth of this favoured lair of the pike. Stretching and weighting the net across the wee loch's outlet channel, we would tie it securely at either end then row off for a night of trout fishing. Monar trout were pink fleshed and due to the abundant feeding at the west end of the loch came fat and tasty. We discovered that they sometimes took well in the disturbed waters churned by the outboard engine, and we occasionally trawled successfully for them in this manner. Mostly however we fished traditionally with rod and fly. Monar trout seemed to prefer quite a large fly and when the fish were really on the take they would sometimes strike two and three together on the same cast. A fine tussle would then commence. The general run of trout over the years were between one and two pounds, with the odd fish up to four or five.

Next morning we'd visit our pike trap and if all had gone well the net lifted heavy with perhaps up to a dozen sizeable fish. Gutting and cleaning the savage mouthed brutes on occasion revealed that their last meal had been an unlucky trout.

The pike being a long narrow fish is a mass of bones but Mrs. MacKay, the boys' mother, had a skilful way of dealing with them and we ended up with tasty fish cakes, not unlike cod. Although dry eating, a dish of April pike made a good meal. The dogs devoured left over fish with healthy relish.

Old Kenny's birthday was the 4th of February. There was always a stir to get him a special present, namely a fresh trout for breakfast. To this end, at the beginning of February, and through March for that matter, we would work the set lines — a simple system employing a hook, a weight and a length of line with float and wooden peg. Then to find the worms. The midden at the byre door or the garden by the house were dug. Worms lay deep but if successful we would make off with the simple equipment to the Strathmore river where it ran deep and sullen between high banks, white with the last season's dead vegetation. Worm baited hooks and lines thrown into the channel at intervals were pegged out on the bank and left floats bobbing gently with the slow current. Next day with luck a float or two might be down and then excitedly I'd pull up a fat trout. These fish at this time of the year took their feed on the muddy river bottom and always seemed in fine condition. The change of diet from the salted venison made a birthday treat indeed.

On and off through the fishing season friends visited us for a day on the loch and the largest pike I saw pulled out of Monar was a twenty seven pounder landed by George Campbell when spinning from the old stone pier at Strathmore. It happened on a sultry day in July when for some reason the pike were running splendidly, and George reeled in around thirty good fish during a couple of hours spinning. Campbell, born in Pait as were his father and grandfather before him, though leaving when a boy, never lost his love of the remote places.

Not to be outdone by way of fishing trips we would often, when gathering over towards Ben Dronaig and taking home old Kenny's sheep from Slis Aillseach (the south side of Meall Mhor), take out rods and tackle. It was mixing business and pleasure. Our route led out through the Gead Lochs, a system of smallish lochans west of Pait with their overspills running into Monar by the Garbh Uisge (the rough waters).

Tramping west on one such expedition our attention was

arrested by the eerie, almost human like shriek of a largish bird. It had risen rapidly from the water beside a tiny island in one of the lochans and with a strong direct flight made off down the glen, long outstretched neck, and strong pointed wings, calling in flight and leaving its most melancholy cry echoing through the hills. We stood puzzled for a moment. I put the glass on the lochan. There on the island sitting about a yard from the edge of the water I recognised a most dignified and uncommon Highland water bird, the Black Throated Diver.

As we approached the lochan, she shuffled into the water in a rather ungainly manner. Once on the water she swam powerfully. Moving swiftly with a graceful sway of her black striped neck and her shapely grey head held watchfully erect, she was obviously agitated. We reached the lochan shore and she dived with a quick over flip, then after what seemed several minutes surfaced on the far side of the lochan. Keen to see her nest I slipped off my clothes and had a cold breath-catching twenty five yard swim. Delightedly I found she was indeed nesting and sitting on two dark brown eggs. They appeared casually placed in a rough scrape amongst the stones, but their sharply pointed ends ensured little chance of rolling out of the hollow. A sharp plunge back to the shore and we left the fine bird in peace, and were satisfied from a distance to see her return to matronly duties.

The divers were often at the Gead Lochs and one spring I recognised and heard the Great Northern Diver. It's amazing 'hallooing' call was quite uncanny and truly fitted these wild parts. There are many legends concerning the Great Diver or 'Loon' as it is called and those who have heard it call will not be surprised. Regrettably it does not nest in the Highlands but farther north. Later that same day we fished west at Loch Calavie without much success but the pleasure of spending placid hours looking across gently lapping waters to the vibrant green and black mass of Ben Dronaig, spun life's thread on timeless wheels.

FILLING THE SALT BARREL

The Red Deer abounded on both Strathmore and Pait and therefore not surprisingly provided us with our year round staple meat. The first fresh venison of the season would be shot in late July or early August. By then the hardened stags' antlers would be cleaned of velvet and the grass taint of the early summer gone from their meat. A young stag or 'knobber' as they are termed would be our aim to start the season with a succulent meat of the very best.

One evening in late July, thoroughly tired of both salt venison and trout, I spied stags on the skyline emerging from Corrie Shaille and grazing quietly east along the face of Creag na Gaoith. Leaving the house with the small .22 rifle and smartly climbing the very steep ground above our home, I stepped quickly along this curving ridge. The ground rose so steeply above the cottage that I used to say you could look down the chimney pots. The wind held steady in direction from the south west, but was increasing. Though out of my sight as I climbed I knew the stags to be in fairly open ground. Well acquainted with the hills' lie I was able to use the curve of the ridge to get rapidly into contact with the unsuspecting group. They came into view ahead and slightly below me. I paused at about two hundred yards and lay still as they hungrily grazed, taking their evening fill. I selected a dark young stag, his two prongs of horns belying his youth. They came step at a time to within seventy yards. I moved to get a better rest. With instant reflex they caught my slightest movement. Heads snapped up. They stood gazing intently straight at me. Without a pause I took the flush of youth through the neck. Down he went, rolling a little with twitching kicks and the steepness of the ground. The fallen pride lodged against a large stone before the carcase could fall further and be damaged. I bled him at once, opening his throat with my knife before dragging the stripling beast down to a more level spot. The first heavy soaking spots of rain fell as I worked at the gralloching. A musky scent of warm blood filled the air as the wind died and allowed the rain to pelt down on my bent back in large drops. My arms stained red as I worked quickly amongst his in'ards, soon white lungs, stomach and entrails lay out on the grass steaming in the rain. Brown–black clouds ended the daylight.

Way below me beside the path which led past the old stone fank, and out to the head of the Strathmore ground, I saw a candle glow light up in the tent of Hugh Fraser. Locally named 'Hughie the Crask' he summered up at Strathmore working on repairs to the hill paths. A man of strength and capability but possessed of a lonely streak he mostly worked and camped by himself. Noting his light I slipped my drag rope round the stag's neck and with a loop in its lower jaw, I hauled the young beast down to the path. Slippery hands cut out part of the liver and walking quietly up to the tent in worsening rain I put my head in at the flap. Hughie lay in blankets, reading by a flickering candle stuck on a jam jar lid.

His great Uncle had vanished in the roaring Yukon Gold Rush of 1898, so Hughie never tired of reading Robert Service and tales of high adventure. From such enthralment he looked up. A sizeable drip narrowly missed the wavering candle. "Is your tent leaking boy?" I enquired. "Leaking be God" he replied, "it wouldna keep out small tatties". Taking the piece of liver from behind my back I threw it into his frying pan which lay handy to one side. Delighted, he rose at once and I left him as he commenced a fry-up.

It was now a black wet night. Dragging the stag to a large stone beside the path, having taken off head and legs to make it as light as possible, I slid it on to my back and walked the couple of miles home in the continuous rain. I need not say that the remainder of the liver went straight into our own frying pan when I reached home.

Filling the salt barrels to keep us in meat through the spring and early summer was a job for January and early February. The hinds, finished with the rut and harvested on the autumn grazings, rose to their peak in body condition, the meat tasting its very best. In the short daylight available much had to be done, especially if the elements turned severe. We would watch the weather affecting the movements of big herds of deer that gathered together at this time of the year. One season, heavy snow had fallen late in January and with the weather holding hard, a large herd of perhaps 400 hinds came stringing down each day to our river flats from Strath Mhuilich. Leaving this high glen away above Strathmore to the north, they trekked the several miles to feed on the still plentiful vegetation about us. This movement involved the herd crossing the ground immediately behind our house. Once the wary hinds checked all was safe they hurried down hungry to graze. Soon the spacious river flats were thickly dotted with ravenous beasts pawing the tussocks clear of snow as with heads unlifting they busily filled themselves before the night frost began to grip.

Noting from Pait the herd's activities the boys came across one night to pass an evening and discuss a plan of action for the morrow. Duly as arranged about midday Kenny rowed over, not without some difficulty as the loch, starting to freeze had a film of cat-ice out from our boathouse. It looked set for hard frost with a brittle bright sky.

In keeping with their practice for several days previously the hinds came down out of Strath Mhuilich at about two o'clock that afternoon. When the last animal of the winding line had settled to feed on the flats, Kenny and I left for the Creag na Gaoith ridge which strategically covered the route along which the deer had just descended. Iain watching from Pait to see us effectively positioned moved into action, crossing snow filled peat hags to the river flats, he pushed towards the herd. The hinds soon had him in full view, a black figure against the white canopy. They trotted a few uneasy steps but with the hard conditions they were reluctant to move from the feeding. However as he approached more closely first one old hind and then the herd took off, hurrying back up the tracks down which they had recently picked their way. Sitting in frosty shadow against a rock on the shoulder of Creag na Gaoith, getting rather cold Kenny and I came instantly alert, as a large dark-red yeld hind led the first of the oncoming herd smartly round the shoulder and up their beaten track a hundred yards below us. With the vigour of the climb their nostrils blew steamy clouds into the chill atmosphere. We sat motionless. The crackling sound of escaping hooves breaking hardening snow told of their haste. Crack. Without hesitation, taking snap aim, Kenny killed the lead animal. A second later my rifle barked in the thin air tumbling another dark coloured hind further back. In immediate confusion they milled about dancing aside from fallen carcases sprawling down the slope. Taking fright they wheeled round, and just as we expected headed back towards the flats. Scrambling to my feet I ran out over the top of the ridge but not without care, in spite of the excitement, as the steep ground was slippery and treacherous. Kenny followed the fleeing deer down the tracks.

Out on the river flats Iain waited. As the deer came running close he swung into action using a heavy old .303 rifle. Coming over the ridge high above the scene I looked down to see Iain at work, his shots splitting the air. A number of deer went down. Kenny with an easy run, came in on their heels. The beasts hesitated. He lay against a stone to kill another good hind. Now it was my turn. The bewildered beasts veered to escape by climbing into Corrie Shaille. Half running, half floundering they struggled now some 200 yards below me. I killed the leading beast. No turning now, they continued running and lunging through the drifts to safety in that high snow blocked corrie. I fired at another large hind far below me. Down she went, only to get up and sprawl downhill wounded. Struggling frantically to escape she flopped and ran in turn. I jumped up and ran slithering down the hill in pursuit. Crossing a heavy trail of blood I could tell she was hard hit. The poor beast dragged herself out onto the flats and as I got close she lay down in exhaustion. I stood over her panting. She lifted her head, eyes stared wildly, bulging, terror stricken. I put the rifle nozzle to her head - click - the magazine was empty. Quickly to my pocket - I had lost the box of cartridges in all the running. Without alternative or hesitation I took my gralloching knife and cutting her throat I killed the gasping creature. The horror of her eye is still with me today.

A most successful drive, we quickly gralloched the kill of nine hinds. Dragging them in to the larder made fast work as they went as easily as a sledge on the crusting snow. Hard frost set in, you felt it tightening your nostrils, sticking them together at each breath. East towards Monar the sky was turning deep purple. The hillside of our day's work stood streaked with bright red trails of freezing blood. After a quick supper of fresh venison liver, we took the paraffin lamps, swinging shadows in the night, down to the larder to skin the kill before it lost all the warmth of life.

Self reliance and self sufficiency became essential and complementary requisites of the life we led. The two or three acres of arable land immediately about the house were provided by the estate for our usage as a perquisite of my shepherd's job. Soon after our arrival I set about with a scythe cutting and clearing the rashes which had crept over some of the ground in preceding years. We were allowed to keep two cows with followers and a pack of 4O breeding ewes. My wages amounted to eight pounds per week minus deductions. That first winter having little other money, we contented ourselves with one cow. She cost thirty six pounds. A few pounds for carriage to Monar and a free walk home up the lochside left us with little excess funds.

That September all carting about the croft was done in a large two wheeled hand barrow. I pulled in the cut and dried rashes to the barn as winter bedding for our first cow. A ton of hay and a few bags of bruised oats were kindly sent by the Estate and 'Margaret' wintered and milked well until about February. In addition to my wage we also received three tons of coal and a ton of potatoes from the Proprietors. Before winter set in these needful supplies arrived by estate lorry at Monar pier. Trying to pick calm days we sailed them up the loch to Strathmore. Though we had no pony ourselves we often had the use of the hardy little favourite 'Dandy' who stayed over at Pait, otherwise it was into the hand barrow for the quarter of a mile pull to the house. Many times however I would carry goods on my back as there was a good sized stone beside the pier at a grand height for the shoulder.

The small garden over the larch fence from the front door after initial turning was a bountiful success. Dressed with unlimited dung from the henshed and byre, it grew the tastiest of greens and carrots. Preserving greens for the winter was a problem. The MacKays who had a much bigger garden of deep peatish soil would bury cabbage hearts in a hole at a good depth in a dry corner. In this way they kept wholesome through to the spring.

Our hens bought in as point of lay pullets we kept in a shed beside the byre, but in the nature of these pestilential flapping afflictions they would be either up at the house doorstep or sneaking off to lay away in the rashes. The children's job was to keep the hens from the door, find their secret nests, and most important, bring the eggs home without dropping them. Any summer laying surplus we put down for winter usage in earthen jars containing water glass, a system which worked well, though my antipathy towards these insufferable creations I never managed to overcome nor explain.

The warmth and hiss of the tilley lamps cheered us all winter. Many might perhaps regard this the biggest nuisance of all, as paraffin is penetrating and smelly, not to mention the meths it always required for lighting. However given a routine and a little care, it was not quite the drawback many might imagine. As I already confessed we rarely, if ever, rose before daylight, so their need was pleasantly limited to the evenings.

Keeping fire and heat however involved a major task. A 'Nu-Rex' stove stood in the kitchen. Hungry maw indeed it served as our main means of cooking and the only source of heat. The remainder of the house, with the exception of a meagre fireplace through in 'the room', remained unheated. During those winters we had some intense frosts, the bed sheets had frozen breath on them in the mornings and one's hand stuck to the metal door knobs. Morning eyes opened to the glisten of frost on the bedroom ceiling. There was a forgiveable inducement to remain well under the blankets. Such conditions certainly highlighted the importance that fuel for heating must have played in more primitive times.

My wood cutting expeditions were an essential operation. At Reidh Cruaidh about a mile and a half down the lochside was a useful stand of pines. They had been planted about eighty years previously to provide shelter for the deer and were of good size. With various storms over the years many had been blown down. These fallen trees were highly prized as I was not expected to cut the timber which sheltered Strathmore Lodge. Lifting my axe I would pull down to Reidh Cruaidh on a post day, anchor at a sandy bay below the wood and start work. Taking the handiest trees I snedded off the branches and cut them into lengths that I could shoulder. Then I picked my way over trunk and stump the fifty yards down to the boat. About twelve or fifteen lengths would fill the boat and I'd load them aboard, pushing her out, after each log, into a little deeper water. The boat deeply loaded I would wade back ashore to wait for a tow home from old Kenny in the *Spray* on his way back from a mail collection. The old rowing boat yawed along behind the launch often in danger of being swamped through my ambitious loads. Landing at Strathmore pier safely and once unloaded I would shoulder the logs home, perhaps not all at once but at any time I happened to pass the pier. It involved plenty of work with the cross-cut saw. Betty and I would pull away at the saw blocking up a couple of logs two or three times a week. Simple steady health inducing exercise every bit as invigorating as the adverts suggest.

One particularly cool bracing autumn day I, in a burst of wellbeing, cut a fair haul of heavy logs and commenced loading the boat. There might have been eight or ten log lengths on board, stem to stern across the seats. I eased the boat out a little to keep her afloat, stepping into the boat as I did to better arrange the timber. A heavy log rolled over, lurching the boat. I slipped backwards. My knees caught the gunwale and down I went backside into the water. The boat heeled smartly over and with her sudden movement the logs rolled down, firmly trapping my legs. In a flash I seized the gunwale with my hands. There I lay, back in the water, legs securely held. Worse, due to the movement, the boat began to drift out into the loch. I wondered how long Kenny would be, how long could my arms hold out? A vision of the postie and old Kenny filling pipes over a good news item flitted unpleasantly into my mind. I had no knowledge of the time. I began to feel decidedly uncomfortable. A measure of apprehension took hold. The boat drifted slowly out and I began to struggle. For some reason that day I had on wellington boots, worn only on the rarest occasions. Holding grimly with my tiring left hand I managed a small movement of the logs with my right arm. Determined, desperate, wriggling and with considerable pain I tore my right leg from its boot. Down into the water, twist round, and I could still reach the bottom with the freed leg. I held the boat, working the other leg clear. In a moment I was free, soaked to the neck but mercifully free. Water up to the waist I pushed the boat ashore. Taking more care I finished the loading. Before too long Kenny sailed up and as he towed me home, I related my recent predicament. As was his way, he seemed unperturbed. "Hoch, hoch, you had a close one that time boy", was his only remark as I began to shiver.

THE LAMBING ROUND

Spring comes late to the high hills of the Highlands. Hard driven winter snow fields are slow to yield the short sweet pastures of lofty corries to the warming May sun, which, nearing its zenith allows scant darkness. Cool winds sometimes speckle the wakening hills with fresh snow, for often bitter sleet showers continue well into June. Overnight frosts and sharp mornings hold back the struggling growth which winter-weary deer and sheep so ardently crave. During these days sharp atmospheric clarity and an intensity of light from the climbing sun present a brilliance of visibility not equalled during any other months of the year. The tender emergence of a fresh season's growth is highlighted in all its subtle shades. The light green and yellow tones of mossy runners which follow hillside water courses contrast against the rich brown ochre of the still dormant heather ridges.

A birch grove below Toll a Chaorachain due to its relative shelter, enjoyed something of a micro-climate. Where more evocatively speaks the spirit of a return to new life than through the gentle, almost imperceptible tingeing of the birches' winter mauve branches with the fairest and lightest shades of leaf green? For this graceful stand of birches hugged the lower reaches of a large corrie. The tumbling waters of a burn to which the corrie gave birth plunged wildly over my secret waterfall hidden deep in this light dappled grove. Its hurried flow was divided for a few yards above the falls by clean grey boulders, only to rejoin in a sparkling cascade, surging over a dark ledge as it showered in a glistening white curve to bubble and churn the surface of a shaded rock bound pool. These carefree waters mixed a melody of tinkling splash and throbbing rumble which played in a timeless harmony against the dripping spray washed walls of patient stone. Velvet moss–clung sentinel boulders stood in angled ranks to guard this shrine of life's elixir. Amongst their clefted pockets, birches and rowan found soil enough to flourish. The trees' rising sap and bursting leaf buds added the first fresh summer scents and turned the shade to a green flickering light. All senses became joyfully absorbant of such atmospheric awakening and this haunt of spring was a favourite resting place when, during May, myself and the dogs would be out and moving on the hills for the long full days of lambing.

With hardy resilience, the breeding ewes are able to survive the bare months of March and April when natural feeding is at its lowest ebb. The sheeps' bodies feel skeleton like to the hand, their wool pulls out in handfuls, yet they live through harsh privation to produce a healthy sturdy lamb. This surely is one of the reassuring features of the shepherd's occupation and often affects his personal philosophy. The will to live is central to the core of existence. Nobody sees this more clearly manifest than the hill man amongst his lambing flock.

Throughout these two hungry months, unless the weather was truly severe, Punch, a large cross Highland garron, would carry for me across his deer saddle three sacks of bruised oats the four miles west along the path to the river flats below the grass withered slopes of Sgurr na Conbhaire. Sighting the pony in the distance was enough to bring the ewes at a bleating run, each from her defined territory on the hill. Hunger rapidly conquers fear. The great mass of the flock would quickly surround myself and pony, eventually to the point where it became difficult to push my way amongst them. Thrusting my crook firmly into the ground, I would tie Punch's reins, and hunching the top bag on to my back, hurry from the path onto the flats, running ahead of the mob to pour out a thin trail of feed from my shoulder. Rid of the milling throng as they ate, I always took a little oats back up to Punch before pouring out the remaining bags. The sheep, never content to eat at one spot ran up and down the feed line pushing and struggling. In a few minutes the meagre rations vanished.

Many old shepherds regarded the winter feeding of ewes as being detrimental to their foraging abilities and consequent hardiness. Latterly I inclined towards the old timers' view. Certainly the flock hung about for hours after being fed, doubtless waiting for more. Neither man nor beast will work if they can avoid it. Once the flock became hooked on hand feeding then the road back to making them self reliant was impossible without loss and death. As the sheep ate I'd select a stone to put my back against, sit on my oilskin and take a careful look about with the indispensible telescope. If all were quiet it was time for a bite of cheese and home baked bread. The humblest feed when eaten out doors becomes a feast. A scrap for the watchful Nancy lying beside me and it was time to move for home. I invariably led the pony. Riding a deer saddle lacked comfort to a degree which made walking truly a pleasure.

Once lambing commenced during the first week of May our feeding routine would cease. The spring growth then took over the flock's nutrition. This avoided a hungry maelstrom of lambing ewes dashing down to the flats which could only but have created mis-mothering problems as greedy mothers left young lambs away on the lambing ground. Newly born lambs call in piteous high tones if deserted for any reason and often unable or unwilling to follow their mothers, they stumble into burns or holes and so perish. This hazard would quickly negate any advantage gained from continued hand feeding. The long days at the lambing took the form of a twice daily routine walk around the lower slopes of the glen. A cut down sack with shoulder strap served as a lambing bag containing medicants for various requirements - oil, penicillin, milk and glucose which, together with my piece, telescope and Nancy were the marching order of the day. These hard and long duties could cover anything up to sixteen miles between the two rounds.

There are few more satisfying experiences for the shepherd than to successfully deliver a living lamb from a ewe suffering the difficulties of a malpresented birth. In order to check that the flock were lambing without problem, strategic stops at vantage points to spy the ewes were frequently required. When all appeared well, with a snap of the glass I'd stroll westward to the next eminence. Spotting a ewe getting to her feet with a long trail of cleansing hanging from her and then turning to lick a shaking little head lifting with its first move of life in the outside world, meant that I could pass by to leave mother and lamb to cement the first bonds of their relationship. Once the bond is made, and this takes only moments, almost all mothers then know instinctively the smell and sound of their offspring amongst hundreds of other lambs no matter how they may be mixed. Occasionally I put the glass on a ewe lying flat on her side, legs out, neck stretched and head forced up as she struggled pressing and heaving to lamb. I'd make myself comfortable behind a stone and watch for ten to twenty minutes. Sometimes I had just spotted a normal lambing, in which case in a short matter of time with one hefty press often accompanied by a loud, long bleat, out would slip the awaited offspring. Having given enough time for the birth to be normal, or perhaps by other signs, I would recognize a case requiring attention.

Approaching a troubled ewe with great stealth, Nancy would be ordered to sit at a little distance from the distressed sheep. Suddenly almost instinctively, the ewe might sense my presence, pause for a second, then up and off at full speed, quite unheeding of her labouring condition. A quick word to the dog "Hold her Nancy", and we'd both tear after the reluctant patient. It's not always easy to catch a bolting ewe on the open hill I assure you. Nancy would run close alongside the animal, grabbing at her neck as they both hurtled downhill. Often no more than a delaying tactic but sufficient to allow me, thundering behind, to lock my crook around her neck and skid to a halt. Nothing pleased the dog more than a successful catch, but she must have been inwardly mortified to be told gruffly to get away and lie down the moment her quarry was in my power. Panting I'd lift the captured ewe and lay her on her side pausing to allow us both to recover our breath. Keeping an outstretched foot on her neck I would then commence to examine the end causing the trouble. Taking my little black lambing bag in a most professional manner, with lubricating oil suitably applied, as gently as possible, I would insert my hand into her passage and commence with sensitive fingering to ascertain what problem lay within that warm slippery world.

Small but strong fingers are the shepherds most useful tool for this delicate operation. Feeling about carefully in order to discover which part of the lamb presented itself was necessary to determine just how the body might be lying in the sheep's uterus. Sometimes your hand came straight upon the nose and mouth of a large head. Here the trouble could be that both front legs were tucked back under the neck, not pointing forwards along with the head as is the normal presentation position. More often it was the head and just one leg that enquiring fingers detected. In either case the head had to be pressed back down the uterus and a finger locked behind the knee of the backward lying leg. A careful pull generally sufficed to bring forward the offending limb to its required position for the correct presentation. All these manipulations were relatively simple.

There were however a number of other complications which could develop. Often just two legs could be felt. This could mean the head was turned backwards or perhaps the rear end of the lamb was towards your hand. Naturally finding the tail gave sound proof! The feet in relation to the body could differentiate these alternatives, and this judgement was critical before any pulling commenced. Whatever the problem, always a degree of tension and concern existed when lambing a ewe. As one carefully manipulated the cosy little form into a position for presentation it would often give a tug of its leg away from your hand. The realisation that achieving a living birth depended upon the shepherd's skill lent much urgency. Taking both front legs between the fingers and giving a steady but not too rapid a pull, in time with the ewe's contractions, generally resulted in a successful delivery. Out would protrude the front legs, followed by the head, which causing the greatest stretching had to be handled gently for fear of tearing the sheep. The body, long and empty, slid easily on to the ground and as the hind legs freed so would burst a rush of lambing waters. There lay a yellow blood stained object which shook its head with the first few spluttering breaths. The navel cord, still attached, would break easily as you pulled the shaking creature round to the prostrate ewe's head. Immediately, a good mother, if not too frightened, commenced licking and cleaning the lamb, at the same time making a quiet low pitched bleat, all in the proudest and most possessive manner. The essential parent — child bonding was forged. Certainly it was not safe to release the ewe until she had begun her task of drying the lamb. If carelessly released without this imperative act she could belt down the hill without a thought for her lamb, and considerably lighter she was virtually uncatchable. You had then another orphan lamb to take back to the kitchen. Betty could already have boxes of lambs close to the fire, much to the childrens' delight I might add. They spent hours playing at dressing the luckless lambs in doll's clothes, giving each a name and expecting impossible tricks to be performed by these cuddlesome bright eyed friends who wriggled bewilderedly as, to the childrens' frustration, they refused to lie with a jersey on in the old pram.

Whatever the nature of lambing difficulties, the saving of a life and the deep pleasure gained from leaving a ewe affectionately tending her new lamb was perhaps the most satisfying experience in all the numerous and varied operations that are connected with sheep husbandry.

Occasionally our ewes succumbed to a condition known as pregnancy toxaemia, a deficiency of calcium and magnesium which left the sheep barely able to stand. Ultimately death ensued if suitable therapy could not be applied.

My lambing round took me across an old crofting site to the east of Strathmore. It was known to us as Hector's croft and was named after Eachan Mhor, a Strathmore shepherd of 140 years ago. The ruins of his little house stood amidst a few acres of once cultivated arable land. Early one bright spring morning I came upon a ewe, heavy in lamb, lying on her side in a semi-conscious condition. I suspected that she was suffering from the toxaemic condition and knew death would not be long away. The sickness was treatable by a large injection of the deficient minerals in solution. This therapy raised the critically low mineral levels in the blood and a rapid cure generally resulted. Unfortunately I had none of this essential curative. The ewe reached out her head, the froth at her mouth told me she was about to die. Casting desperately about I wondered what might be done. Unquestionably the ewe would die, and with her, the unborn lamb. With quick resolve, I grabbed a heavy stone. One swinging blow to the ewe's head killed her. Blood spurted from her nose, eyes took an immediate unmoving stare, and though choking breaths bubbled out blood I knew she felt nothing. Using my skinning knife and working with feverish speed I opened her stomach and cut into the womb. Fluid poured out. Curled inside a thin yellow membrane sack snuggled a heavy lamb. To my great joy it came alive in my hands as I slit open this last barrier. At once I carried him, for it was a large tup lamb, home in my lambing bag and put him straight in front of the fire. After he dried out, a drop of milk through a baby's teat started him feeding. He lived and throve, soon running about the kitchen with gusto.

After a week I found him a ewe which due to bad weather had lost her own lamb. We set him on to his adopted mother in the old fashioned way. The skin from the dead lamb he was replacing had to form a little suit for our orphan. The knack, for there is always a knack, lay in clever skinning. The dead lamb's skin was removed so that the neck hole and the four leg holes remained much as the sleeves of a jersey. Skinning only took a few minutes then Rupert, as he had been christened by the children, was pushed squirming and wriggling into his new jacket. First over his head, and then each foot into its sleeve and with finally a smear of the new mother's lambing waters on his head, he was ready for her inspection. Holding the ewe with my knee whilst pushing him down to her udder, I saw he lost no time in nosing under her flank in search of a teat. Once Rupert was strongly sucking I released the ewe. Quickly she turned her head to his tail and sniffed the skin. I had made sure that the jacket came well down over the lamb's tail to be sure of hiding his original smell. After several testing sniffs she stamped a foot at me. This protective action to my great relief indicated the ewe had taken him as her own lost lamb without any doubts. They spent three days running about the croft before I freed Rupert from his now rather smelly jacket. Later that morning, I set them both away out of the west gate to the open hill. The lucky chap trotted in to the ewe's side quite the thing and didn't look back.

Cruel weather during these weeks created much work for myself and the dogs. High in the sleet swept glen a single bad night could reduce the lambing percentage drastically with a trail of dead lambs found the next morning stretched out flat and formless. In a short time the little bodies quickly had their eyes and entrails pecked out by the scavenging 'Hoodies'. One felt so helpless in the face of such seemingly demoniac elements and raging I openly cursed the weather for its blind indifference. Nevertheless during the fine spells, should lambing be going along without incident, ample time afforded many hours of pleasure in studying the varied wild life all about me, busying itself in spring-time fancies leading to the discipline of catering for the next generation.

Lapwings were invariably the first of the new season's visitors to inspect their breeding ground. I recorded some arrivals as early as January 28th. During an open spell of weather large flocks would appear from the east for an excited check of the territory. The calling flocks of birds came in waves which seemed to undulate as they drifted in with heavy wing beats, their black and white markings in stark contrast to the grey February sky. They gave us the first welcome cry of spring. After several journeys between us and the coast the flocks paired off and settled in March to nest throughout the extensive rashes about the croft and the river mouth. One trusting pair nested not fifteen yards from our stick shed and during courtship this friendly couple exhibited their excited aerobatic tumble and cry over our heads at frequent intervals. From such close quarters as they ran to and fro, the strong morning light would catch a wing plumage which I can only liken to dark Mother of Pearl. At each turn it flashed a colour quite unsuspected, browns and greens, mauve to purple, all seen through a silk-like sheen. They reared successfully and I was careful to keep the dogs from snatching up the helpless chicks that scurried about the rashes for many days before being taken further afield to safety by continually anxious parents.

Not to be outplaced in the rush for nesting sites, the Redshank and Oystercatcher flew in, as I recorded one year, on the 9th of February. Most watchful of all the waders, the Redshank with its rapid flight and Te-tu-tu-tu alarm call nested at the foot of the croft. To my slight annoyance one handsome male Redshank took the ridge of our cottage as his observation tower. Each morning at a dreamingly reposeful hour, around 3.30 a.m. he would sound reveille. The third morning of this impudent intrusion, I leapt out of bed and stuck my head through the skylight. At two yards our eyes met. His, bright, black and orange, alert and challenging, mine rage filled and vengeful, I sought to wither him at a glance. The confrontation lasted some seconds before he darted away, white rump and rapid wing beats. In twenty minutes however he was back. The bed was warm, I conceded defeat and so he continued to address the locality from our rooftop for some unmolested weeks until his brood down at the croft fence was safely reared.

Most evocative of the spirit of spring is the hauntingly primitive call of the curlew. Never early to arrive, they flew to us on powerful wing beats perhaps late in April. On arrival we only received a shortened version of their magnificent song, but as the long mild days of May came in and nesting territories were defined, the glen walls rang and echoed to this pure essence of the wild Highlands. "Curlew, curlew" the song started slowly and quietly as if at great distance, then swelling, filling and flowing, this untameable music would rise in pitch and tempo to burst in loud and plaintive crescendo across the ageless hills, filling each corrie with a melancholy which seemed to reach back to their very creation.

A notably rare species which nested with us was the Greenshank. It could be described as not unlike a cross between the Redshank and Curlew. The name correctly suggests the legs long and green, its bill unlike the Curlew is slightly upturned. Only once did I find the nest of this exceptionally shy bird, and that by accident when one of the dogs put up a sitting bird. She was incubating four light buff eggs, beautiful with dark markings. This pair had nested at a little distance from their marshy feeding ground which was the Strathmore rivermouth and regularly commuted to these flats each morning and evening during the incubating period.

One spring an ex-army 'Gentleman' stayed at Strathmore. A writer of note on natural history and like topics, his obsession was to find and report to the general public upon this elusive bird. We quickly christened him 'Colonel Greenshank' as he talked of little else. Unfortunately his ornithological investigations were cut short before many days. Busying himself in flattening with his boot a rusty nail discovered protruding through a piece of flotsam at the edge of the little loch's water line it rather to his surprise, stuck smartly into his foot. A major rescue operation was mounted. As I boated him down to Monar, he sat with a troubled look and his foot up on the seat. The Monar keeper obligingly ran him the thirty miles to a doctor whilst I awaited his return. Limping and suitably bandaged he reappeared some hours later and we sailed back to Strathmore. He proceeded then to direct us in a search for Greenshanks from a wicker chair in the Lodge. Meantime I was given the blame for his misfortune and a letter to this effect he promptly dispatched to my employer. After ten days he recovered and resumed active command.

I had much respect for Colonel Greenshank's extensive knowledge of ornithology. Taking him along the hump backed top of Meall Mhor one clear day early in June, we spied a small bird which was certainly strange to me. As it flitted about in the shattered ridge stones seemingly unafraid we got a close look. Greatly excited he identified the bird as a Snow Bunting in summer breeding plumage and we watched to ascertain whether or not it might be nesting. After some minutes the little golden coloured bird flew up the ridge with a jerky pausing manner. The 'Colonel' verified his observation and in animated conversation informed me of the importance of our day. This species of Bunting normally nesting in the Arctic Tundra had been suspected, but never proven, to breed in Scotland. Conspicuously ignoring any duties I might have to perform for my employer, I was dispatched the following day to check the site of our encounter with the Snow Bunting. Rather to my own disappointment and greatly to that of the 'Colonel Greenshank' I did not see the bird during a hot trudge the length of Meall Mhor.

Certainly hundreds and perhaps thousands of waders used the Strathmore flats as their breeding ground. The sound of their calling, courting, feeding and sometimes alarm notes ended the heavy silence of winter and gave us six weeks full of unrivalled song. The snipe displaying over bog and swamp below the Lodge could still be drumming in the soft twilight of a May midnight, an eerie sound which frightened the children. A couple of hours lull in the semi-darkness gave way to a chorus of awakening life as each pair went about their early feeding. Early on a breathless sun bright morning the fingered yellow rays of growing light were richly accompanied by this natural orchestra.

I spent one interesting day photographing a breeding Dunlin. This chubby little wader, golden brown back and black breast, I found nesting on the mud flats where the meandering river entered the loch. It ran about agitatedly on the mud just a foot or two from me as I lay on the bank taking snaps, in what I suspected to be a somewhat unprofessional manner. An early hatch of Mallard swam about in the pool that day but failed to appear as more than minute specks on the prints. My aspirations to become a wild life photographer were realistically tempered. Mallards were common nesters in the area, whereas the little Teal duck seemed less so. This latter russet headed duck did however nest in the area and some days later, walking in deep heather on the lower slopes of Meall Mhor at least half a mile from the loch, the dogs put up a female Teal. She had risen from a tunnel which led down two feet through the thickest of heather to a beautiful bowl shaped nest made of dead grass and lined with her thick dark down. There lay nine creamy eggs. The dogs having scented her out were keen for a quick snack but a sharp word prevented this happening. The Widgeon, another beautifully auburn headed duck, I knew by its long whistling musical call, which sounded cheerfully from the marshes. I never found their nests, nor did I make it my business to search out nesting sites, but in an area so profusely used, seldom a spring day passed without the pleasure of finding the birds' varied and beautiful homes.

Hovering over the busy scene at certain times of the day would be the Kestrel which I knew nested and reared on a cliff face above Toll a' Chaorachain. Its appearance over the summer flats was always accompanied by the screeching and mobbing of intrepid little Pipits, often half a dozen or more flitting in close like schoolboys daring a bully. The buzzards that nested straight above us in the rocks of Creag na Gaoithe also suffered the same daring darting rushes by anxious nesters. From our croft, with the sharp morning sun detailing the black cliffs of Creag na Gaoithe, I witnessed the buzzards not far from their accustomed nesting site, attacking a Raven.

This ominously solitary bird seldom came down as far as our relative civilisation, preferring the truly remote areas. Jet black and massive, the nature of its lordly flight gave reason for respect if not apprehension. This day the black master was passing on business best known to himself when to his great indignity he was set upon by our pair of Buzzards. I was scything hay at the back of the house when the cat like screaming of the latter birds drew my eye up to the rocks. I ran for the glass to witness a battle royal. The Raven, shining black and a good deal larger than the Buzzards clearly displayed the better wing power. The Buzzards mobbed in close to collision before he would turn and drop, only to soar away on an up-current and dodge his assailants. The fight lasted with great ferocity, for some ten minutes before the Raven, in a final great wing closed drop then rose powerfully to soar away, and with strong wing beats made back to the privacy of his far glen. I wondered at the encounter, but knew that the Buzzards nest often had scraps of rabbit lying about. I had crawled out to a dangerous ledge and looked down on it a few days previously. The two fluffy chicks lay sleeping in the stick nest and had food debris scattered around them. Perhaps the passing Raven had been attracted by the smell.

Thus the lambing season stretched on with each day affording many hours of interest in all the bustling life about me, enjoying the deepening summer green of Strathmore glen. In the hot sun one June afternoon, I took off my shirt and lay on the grassy bank beside the happily gurgling river. Though half asleep my ears caught every slight sound and a 'plop' like a small pebble falling into the water took my attention. Turning over from the warm sunlight I looked down into the clear light brown waters. There swimming under the surface, quite rapidly, was something which gave the effect of a shooting silvery bullet. Reaching the far bank of this little enclosed backwater, out scampered a small black animal with a short nose and stumpy tail. It sat up and pawed its whiskers in an amusingly human manner. Then to my surprise from a hole low down in the bank out scampered the remainder of this family of water voles. They scuttled about on the mud emitting sharp, high-pitched squeaky noises as though excited that Father was home. The first vole I had noticed plopped back in to the clear waters and the family dutifully followed. I saw that the silver effect was due to the bright sun glistening on the air bubbles which seemed attached to their sleek black coats. They frolicked and dived about the pool for all the world like a sports day at school. The sun grew warm on my back, I lived with the little voles in a carefree world.

Later that afternoon the scene was to change. The incessant calling of a ewe at the foot of Corrie Sheasgaich came drifting across to me. From the tone of her bleat I knew she had lost a lamb, so pulling on my shirt I set off up the face to investigate. The ewe was bleating as she ate in the mournful way they do, but on my approach ran over a few yards to stand beside a dead lamb calling distractedly for it to move. She bolted as I picked up the carcase for examination. The lamb's neck showed clear teeth marks and had obviously been killed that morning. I knew this to be the work of a fox and with a thought put the body under a stone. Nothing could be done immediately but I went over to Pait that evening to discuss the matter with the MacKays. Yes, I had seen plenty of fox droppings in the glen, particularly on the large tussocks which served as their signalling posts beside the path. Frequently the extremely pungent scent of the fox came to my nose as I crossed their tracks, so strong a smell that it was detectable for hours after the fox had passed. All in all the boys thought an investigation of the big cairns of Sgurr na Conbhaire might be the first move and duly arranged to come across the following morning.

Leading the hunters up the path from the pier next day came Spirach, the MacKays' terrier. The Gaelic name means Sparrow and indeed she was quite small. Tufts of hair hung over, almost hiding keen sharp eyes; tiny stand-up ears were followed by a roly-poly body all supported on what seemed no more than three inch legs. For fierceness she knew no match and Iain kept her up to scratch by constant teasing, much I might say, to his Mother's displeasure, who regarded the dog as something of a pet.

Each with a gun, Kenny in addition carrying a rifle, we tramped up the glen and past the old stone fank at the foot of Toll a' Chaorachain Spirach in high glee darted about, paddling little legs and quivering tail all going at a great rate. As it was an east wind, we kept along the glen path until we were well to the west side of the cairn. This impressive jumble of the hill's fallen masonry, stone piled on stone, was a veritable warren of holes and tunnels. After a careful spy we approached the huge cairn silently, crossing the gullies and small water courses which lay to its west side. Spirach now on a lead and pulling frantically, indicated we could be on to an adventurous trail.

Tension increased caution we stepped quietly towards the main area of tunnels. A nod, no word, and Kenny with the rifle moved out above us. Iain and I made on stealthily. Twenty yards from the den, out between the boulders, flashed a long russet body running hard up the cairn. I leapt on to a stone. "Kenny, Kenny, he's coming up to you, look out". Recklessly jumping and dashing over leg-breaking boulders we flew the last few yards to the fox hole. Spirach went wild, leaping into the air and biting her chain. Above us, crack . . a shot followed by a ricocheting whine; pause and then another crack. Iain stayed at the den, but I leapt up the stones calling as I ran, "Did you get him Kenny?" "Yes, yes", Kenny's excited voice floated down. Unheeding of breaking legs amongst the moonscape rocks, I clambered up to where Kenny, grinning evilly, was now tying a string to the hind legs of a large dog fox.

What a fine fellow the fox had been, a deep russet coat against the clean cream of his underparts, a tail eighteen inches long, fat and broad to its large white tip. His mouth, half open, held large sound molars and fangs, bespeaking the prime of middle age. A trickle of blood oozed over his tongue for Kenny had made a clean second shot through the chest. He had died instantly as his wide bright yellow eyes told me. We made down to Iain, pulling our bumping trophy by a string. By now he had Spirach chained at the mouth of the hole. We held council. If the vixen lay in the den, which now seemed likely, then we could take more time. The den was very deep and, so far as the boys knew from past visits, had no means of egress except by the hole which we now guarded.

We sat and took our piece, admiring at the same time the handsome dog fox. Due to his pungent smell, we left the limp form some yards away. Angled boulders cut into clean lines the warm sun that fell into the tiny moss floored dell in which we ate. The area showed all the signs of fox activity, bits of bone and skin lay strewn around. A number of lambs had obviously been killed and carried in to feed the litter of cubs we guessed would be hiding apprehensively underground.

Cheese and oatcake finished, now was time for action. Loading our 12 bore guns, Kenny and myself took up commanding positions of rocks either side above the entrance hole. Waiting until we were in position, Iain released the still straining terrier. Straight down the hole she vanished and Iain smartly climbed another stone overlooking the den. We covered the ground, fingers feeling for triggers. Spirach's shrill barks, loud at first became fainter and fainter. "It's a very deep cairn", Iain spoke up to us. We waited tense, guns cocked, minutes passing slowly. Without warning, unhurried, keeping close to the ground, the vixen slunk out. She paused momentarily at her dead mate. Our guns roared out in unison. Over and over she bounced from the combined impact. Tension high, we waited to see if any cubs would bolt, but after some little time Iain looked up and said quietly, "nothing likely to emerge now".

We stepped down into the arena. Faintly from the tunnel mouth came frenzied barking, indicating that Spirach was engaged in a vicious fight with the cubs. Still excited we shouted down encouragement to the battle in the cairn but after a while just sat waiting for the terrier to emerge. The evening drew in. All barking had long stopped. Finally we could wait no longer and cutting the tails from the two adult bodies, as proof of the kill, we buried the pair under stones and left for home minus the heroine of the day.

The foxes dearly paid the price of their damage, and so it seemed had we, by the loss of Spirach. The welcome meal Betty had ready we ate with small comment on the day. Needless to say Mrs MacKay was both annoyed and upset. Days went by. Spirach was truly given up for lost. No doubt killed in the cairn by strong cubs. Never mind, through the night eight days after the successful destruction of the marauding foxes, a scuffling and tiny barking awakened me. Running downstairs and opening the porch door there danced an emaciated, woebegone Spirach looking up at me, madly wagging her tail. It was a fine night. I rowed her over to Pait there and then. She leapt out of the boat at the pier and shot off up the track. Her homecoming barks wakened the whole household. She obviously had been stuck in the cairn, due perhaps to forcing herself through a narrow hole during all the fighting. From this morbid incarceration only starvation could yield freedom.

The local Fox Destruction Club, a body set up to help sheep farmers protect their flocks against the consequence of a numerous fox population, covered a large area. A full time fox hunter employed by the club came and stayed with the MacKays at certain times of the year whilst undertaking his duties. Late one spring Johnnie MacRae the club's hunter rowed over from his digs for a routine round of the fox dens on Strathmore. I invariably accompanied him, both for the pleasure of his company, and to carry some of the equipment. Two hardy terriers were his companions.

We set out on a day of lifting mists for the cold remote corrie of Toll a' Choin which lay north east of Strathmore. Surrounded by precipitous crags and at the foot of the vast bulk of Maoile Lunndaidh, this sunless corrie seen in creeping mist called up a foreboding which spoke deep inside one's mind. We entered the corrie from the south and with the wind putting down, the dogs became very agitated at least half a mile from our destination. Here was a scree of primeval appearance and proportions, a vast conglomeration of angular boulders carelessly strewn down the steep hill slope, having crashed in a massive rock slip from the cliff above. As Johnnie remarked drily at nature's act of vandalism "Indeed you would be the better of a tin helmet if you were passing the day that lot came down".

Making as few sound-carrying metallic clicks as possible we loaded rifle and gun and approached the fox holes slowly. An eerie, silent mist moved like wispy stage curtains showing crag and gully for a moment then enveloping each changing set in billowy white drapes. Too late I caught the flash of a large fox, which I presumed the dog, making fast over the sky line. Primitive, atavistic, the lust for the hunt drove us breathless up to the den, where Johnnie let the terriers loose. They ran excitedly about the numerous holes, finally deciding upon the most likely. Taut, keyed like a coiled spring we watched, eyes shining to kill. Bronyan went to ground. Seconds thumped in my throat, senses electric, eyes sharp, darting towards any movement. Then out sped the vixen a terrier at her throat. Brochka joined the running fight. Unleashed I flew over the rocks feet on wings, trying to keep the hunt in sight. Down, down we dashed, blood-maddened man and dogs, heedless one slip would split my head. At the bottom of the cairn I leapt into a hollow beside the incensed, frenzied blood-lusting terriers, growling shaking and tearing the already dead vixen. The innermost disappointment of not killing I cannot deny nor explain.

I caught the fox by a leg, shaking and kicking off the furies, to drag her back up to Johnnie who was watching the den holes, gun at the ready. Like little maniacs they worried the now bloody torn body and I cursed them roundly, perhaps covering my frustration. At Johnnie's word the dogs stopped swinging on the dead fox and ran about looking for more work. The vixen we noted from her pale udder and nipples now bleeding and ripped, was heavy in milk. The young cubs would be in the den. Hardly needing the encouraging noises the fox hunter made, the dogs flew straight to work scrambling, sniffing, squeezing into holes, backing out and trying the next, little eyes blazing still with killer instinct. In a few seconds down into another likely looking hole they vanished, where much barking and squealing could be faintly heard. Suddenly Brochka appeared from the hole dragging a living cub. Johnnie in a flash seized it from the terrier's mouth and put the terror-rigid cub straight into his keeper's bag. The surprised terrier turned about and disappeared instantly down the same hole. Two more cubs were carried out, limp, crumpled, downy bodies both dead. After some little time snuffling and scrapping the bloodstained and triumphant dogs pleased with themselves, knew their day of destruction was ended. They settled to lying and licking whilst we took up the task of setting the gin traps, which I had carried up, at various chosen points on the pathways into the lair. This strategy was undertaken in the hope that by evening the dog would return to the scene and in his distress be caught in one of the traps. This aspect did not appeal to me. The afternoon, lifting to a soft breeze from the west, drew on. Blinks of cloud-beleaguered sun startled the forlorn corrie from its reverie upon misfortune.

Carrying the cub we set off home looking forward to a large salt venison supper down at Strathmore. Coming into the house Johnnié, who was very fond of the children, unslung his keepers bag and with a great air of mystery set it on the kitchen floor. By and by Hector, when he thought the keeper not looking, peeped in, only to start back from two bright little eyes staring out at him. We'd all been watching surreptitiously and burst out laughing at the boy's tearful discomfort. The cub was let out and ran whining about the room, looking round with wondering eyes, scuttling under chairs. It was the bonniest bundle the children had ever seen. Its sharp intelligent face and dark fluffy coat made it seem an ideal pet. They scrambled after the moving toy in high glee. However after supper that evening we had a job for the cub and set off back to the same corrie.

Both this time with rifles, we approached the den in the evening's long shadowed sunlight to find to my private relief, for I detested the gin trap, that they remained unsprung. Picking a dry position, offering wide views, we settled down to await in failing light a chance of the dog fox returning. It grew cold, we made no talk and waited. A cackling grouse somewhere above us brought a meaning look across to me from the hunter. "Was the dog moving in?" his look said to me. I looked east. The ground swept away from the corrielip. Undulations of hillock and peat hag, a rich yellow warmth of dark folding shadow as the trailing sun took each feature in an evening ritual of blessing with light. Lacing burns picked their way round and down to find rest in Monar's wide mountain locked waters. Breaking the smooth silhouette of Maoile Choill-Mhias, a herd of hinds grazed in step. Another month would find them spreading to high ground seeking seclusion for calving.

Out beyond us the Strathfarrar hills thoughtful in sombre meditation looked especially high and close that night giving sure warning of rain by morning. Was our mission that day, as it now seemed, a wanton intrusion? In such reflection the gloaming remorsefully closed round us.

Time for action though, the cub was taken out of the bag and he began to squeak loudly, not least as his ears were being pinched. The piercing treble carried through the gloom. Almost at once I saw the dog sit up on the skyline. He had clearly been watching our movements but lying low until this sound of distress caused him to abandon caution. Cub forgotten, Johnnie grabbed his .22 stalking rifle and using the telescopic sight due to the poor light, took a splendid long shot which tumbled the fox down into the cairn. Noting the spot as best I could, I ran up in the darkening and after casting about, found the faithful animal lying dead. A clean shot, I wasted no time in cutting off his brush and throwing him under a stone before hurrying down to Johnnie, who by good luck had caught the simpering cub as it ran over to its dead mother. We gathered up the traps and picked our way home through a night now black with threatening rain. To the childrens' disappointment Johnnie, terriers and cub went away down the loch the next day.

ON THE LOCH

"Keep her nose into the wind, keep her nose into the wind". Kenny 'Tin', the old Strathmore shepherd swayed bravely about in the hold of the *Spray* which was violently pitching and tossing its way west up Loch Monar into heavy foam topped breakers. White backed they creamed past the boat driven by a shrieking gale, which when funnelled through the hills, added a force which lifted crumbling wave tops into smoke like trails across the heaving surface of the loch. The rollers towered above the boat as she climbed out through each curling crest to crash down upon the back of the next wave with a mighty thud which sent spray whipping into the face of the man at the tiller. A rough day for Iain to be escorting home to Strathmore my shepherding predecessor.

This seasoned bar counter compaigner was returning from a foray 'down country'. Such periodic events enlivened the life of this old bachelor shepherd and, not least I might add, those with whom he came into contact. His dog 'Prince' cowered amongst the tarpaulins which lay tucked under the decking at the bow. Tail tucked between legs he was quietly being sick, partly as a result of the extreme conditions on board the boat but more particularly perhaps due to his owner's unpredictable behaviour. Kenny 'Tin' wearing a heavy harris tweed greatcoat over his smart plus four knicker suit, staggered from gunwale to gunwale miraculously keeping his feet in spite of the rapidly changing angle of the deck boards. At intervals he ducked under the cabin, sat on the bench and from a half whisky bottle found in his pocket drew the courage he clearly felt the severity of the occasion demanded. Suitably fortified he would re-emerge with explicit and a trifle impolite instructions to the helmsmen. "Sit down, sit down", yelled MacKay over the tearing wind. Kenny, ignoring such faint hearted advice turned to face the elements which by now would be presenting a daunting sight to all but those of the stoutest heart.

His rapidly assumed seafaring eye caught sight of a truly magnificent breaker, three waves ahead, bearing ominously down on the pitching craft. Gesticulating wildly towards the oncoming peril, he turned with a look of nautical concern somewhat undermined by a certain lack of naval dignity and screamed, "keep her nose into the wind, Mackay, into the wind you stupid b....r! *I've* sailed round Kilchoan point" – the latter fact being added by way of impressing upon a landlubbing helmsman the breadth of his own seafaring knowledge and ability. "You'll go overboard you old fool" Iain bellowed cuphanded. The warning barely left the helmsman's mouth when with a whoosh the full lift of the feared wave gripped the *Spray*. Like a rearing horse she suddenly tossed her bow dangerously heavenwards and following a neat trajectory, Kenny catapulted through the air with graceful ease straight into the loch. This abruptly cut off the flow of pilot instructions.

Never have I known anyone as quick of action as MacKay. It showed to Kenny's benefit in an instant. Throwing over the tiller at alarming risk, he put the boat broadside to the waves. That the *Spray* didn't turn turtle remains a tribute to her lifeboat design. The man overboard, floating thanks to the spread of his greatcoat, but blowing water piteously and waving dementedly, was passing under the lee of the boat. A hand shot out and grabbed a lapel, a foot spun the engine out of gear, in case Kenny's feet touched the powerfully turning propeller MacKay clung on. The tackets of the shepherd's hill shoes he heard dinging on the brass propeller. A decidedly startled face looked up. Iain leaned out watching for the next wave. A roll of the boat and Iain took the most dangerous step of all. He left the engine cockpit, stepped out onto the open deck and as the *Spray* dipped her gunwale, hauled the spluttering Kenny back into the hold. Dropping the soaking human bundle, MacKay leapt back into the cockpit put the *Spray* safely under power and swung her back into the wind.

The seaman cum shepherd lay for some well soaked minutes before bestirring. He appeared undismayed and in no way contrite. Using the security of all fours he regained his seat. With memory contractions common to those considerably over indulged, he felt the event merited further sustenance. Shortly the empty bottle flew past MacKay's head.

The *Spray* ploughed on. The seaman-shepherd evinced recovery signs and now thoroughly disgusted at the lack of helmsmanship, got up again, oblivious of his soaking clothes or recent escape. In an athletic leap Iain jumped out of the cockpit into the hold and delivered the hero a stiff punch to the side of the head. He fell to the boards and lay still. Water ran from him into the bilges and back again. A hand moved limply to another pocket which contained the remains from a previous feast. There he sprawled, eyeing MacKay maliciously around the sides of a bottle, until finally his grip slackened. The bottle slipped to the boards. The last of its golden contents trickled into the bilges and Kenny, known as 'The Tin' fell asleep.

Thankfully arriving at Strathmore pier, Iain moored the sturdy craft. Catching slumbering Kenny below the oxters he lugged him out of the boat and set him on a large stone. Heaving the soggy bundle across his shoulders Iain carried the oblivious shepherd up to the cottage. Kenny didn't waken. Iain lit a fire and left the heap to steam lying in a chair and caring for himself. Next day the weather moderated as quickly as it had risen to storm, the loch subsided into calm reflections. Kenny was spotted rowing across to Pait about dinner time. He appeared at the door hale and hearty. Mrs MacKay who knew of the escapade said nothing, busying herself giving him a plate of soup, but after some little time innocently observed "Oh Kenny, what a bruise you have on your cheek". "Is that so?", replied the unrepentant survivor. Like a true Highlander he said no more, and neither did 'Teenie'.

Boat work was a daily part of life at the west end of Loch Monar. There were two motor boats, one the *Spray*, already described, and a second smaller launch, the carvel built *An Gead*. The latter was named after the pleasant hill lochs which lay to the west of Pait. Both boats were powered by petrol-paraffin, four and two cylinder engines respectively, which, if cared for properly, gave the minimum of trouble. The setting and rake of the *Spray* engine did however lead to some problems in its lubricating system. This was modified by lowering the engine on its fittings, thus bringing the engine propeller shaft axis nearly parallel with the keel line. For this operation she was taken down to Beauly on a low loading trailer. During the *Spray's* alterations we depended upon *An Gead*, which though a sound boat to handle, was in what can only be described a 'ripe' condition and as a consequence gave us some nervous moments.

The Pait pier and boathouse, had been well constructed in stone and concrete, and the facilities ran out into deep water. On the other hand, Strathmore pier, which lay directly across half a mile of the loch's headwaters from the Pait landing, was less suitable for the launches and being a shallower approach only a wooden boathouse and natural stone pier were deemed necessary. Pait stalker old Kenny MacKay, the two boys' father, grandly styled himself official postman and perforce became boatman engineer. As an employee of the G.P.O., he had solemnly signed the Official Secrets Act, cheerfully accepted regulation issue number, cap and bag but more significantly a state emolument to collect and deliver his own and our mails on Tuesdays, Thursdays and Saturdays. A finer, pleasanter and more entertaining old man it has never been my pleasure to meet. Kenny in nature typified that imperturbability common to people who live far from the modern life and at peace with their surroundings. He certainly knew no fear as far as the loch was concerned as this next little incident implies.

Kenny oilskinned, casual but competent, sailed out in the old *An Gead* taking his employer and wife down to Monar on a day of particularly savage easterly gale. Sir John, knowing well the elements, had hesitated about setting forth but due to an important meeting decided upon the risk. Himself and Lady Stirling sat huddled in the bow of the boat preferring at least not to anticipate doom by looking out. After clearing the sand bar the game old vessel was struck by an exceptional wave. Sir John admitted "I thought we were all goners, but I looked up to see old Kenny calmly lighting his pipe and guessed then we might still survive".

The motor launches lay at Pait in old Kenny's charge along with several rowing boats, whilst I kept a couple of ten and twelve foot row boats at Strathmore. Old Kenny often passed a day down at the boats, working on the engines and general maintainance. Naturally the half mile walk to the pier from his house was a routine which gave him no thought, but on occasions there might be something heavy to carry and he would take down the jeep. This vehicle served us faithfully, for all we knew it had been through the desert war. Kenny, an ex Scots

Guards' piper of first world war vintage, treated it as a comrade in need. Thinking back to the trenches he'd put the 'buggy' into the garage with a remark "well boy, you'll be the better of a dry bed tonight. Manys the night I was wishing I had one so good". Alas it happened this first Pait jeep became beyond old Kenny's affections and mechanical skill.

One calm day we loaded the vehicle onto the *Spray* and shipped it down to Monar. A dangerous and top heavy load which we were relieved to drive off its planks and up to the Monar pier. Incredibly in its place we received a second jeep of even older vintage, equipped with certain sporting features, notably an unpredictable gear box. It imparted a distinct thrill of anticipation to be never quite clear which gear the contraption had in mind to engage, but once engaged became even less clear when a decision to disengage might be taken. In short, there seemed little connection between what the machine did and what one wanted it to do.

One hot summer's afternoon, the MacKay boys and myself idled on the Pait pier blethering away as we waited for old Kenny coming down from the house by means of this singular contraption. How easy our conversation, there was no hurry to say anything. During long pauses the sun fell friendly on upturned faces. The whole summer had been sunny and dry and the loch stood very low. Boats swung gently, moored to the end of the pier. Blue powdered hills lay about us, their reflections creased into rippled smiles, for a mischeivous breeze played tig amongst the happy reed beds. The silence droned elusive, yet captivating, an almost touchable melody of peaceful content which hides in remote worlds far away and long ago. Eyelids fell, seduced by a dreamful harmony, hours could be minutes, for time slept in the heat.

Stirring reluctantly our attention was taken by the sound of the approaching jeep. I casually remarked as we listened "I'll tell you this boys, he's coming down at a fair toot". Indeed the engine sounded almost to be screaming with pain and without doubt was approaching with considerable rapidity. Iain rose from the dyke upon which we sat sunning and strolled to the corner above the boathouse. "Look out", he yelled in the nick of time. Old Kenny tore round the last bend to the pier, gravel flew, engine roared. We leapt for our lives into the boat. The crouching postman–cum–engineer with a maniacal daring befitting only a suicidal stunt driver, careered down the pier and shot straight off the end. With a mighty splash he planed out across the water. First to our horror and then our amazement, he continued to drive the crazed machine out in to the loch. Gaining control or perhaps never losing it, Kenny heroically swung round the amphibious monster before it totally submerged and headed back inshore. Together man and machine climbed a steep bank, bumped back onto the track and headed, without any reduction of speed, up the road they had just descended. The engine still in hysterics roared away.

We clambered out of our refuge and listened. "Only a wall will stop it" reasoned young Kenny logically, though showing some small concern for his aged Father as the sound carried him on up the road. "He might have given us a wave in passing" I ventured a trifle unfeelingly. Mesmerized we watched the jeep tear into the yard, stunned to inaction as we waited for the crash. Suddenly all fell quiet. Flies started to buzz on the warm boathouse wall. Fifteen minutes later to our immense relief old Kenny walked down to join us at the pier. No trace of excitement seemed evident on his weather beaten face. "Isn't it lucky the loch is down boys", was the closest he came in a reference to the antics of the brute he had just abandoned. Only later did we discover that the throttle had stuck full open. The jeep for reasons best known to itself had mercifully run out of petrol just as it carried its apprehensive driver into the yard. "Indeed I thought I was for another turn down to the pier. The bitch would neither switch off nor out of gear would she come", he laughed when recalling the event from beside the peat fire.

On several occasions I rowed up and down to Monar, always choosing a calm day of course. Given a following wind the pull would take something over an hour and was not too strenuous. Latterly I fitted a primitive square sail to one of the rowing boats and sailed down the loch in fine style steering with an oar. These journeys served to increase my admiration for the toughness of the old timers before the advent of motor boats. Sir John's father commented upon the hardiness of the ghillies who at the end of a day's stalking on the Strathmore hills would set off and row down to Monar. The sturdy boats were manned by four oars, always with a steersman and room for passengers with luggage.

In the 1830's, living at Luib an Inbhir, half way down the north side of the loch was a family by the name of MacPhail. They were employed at that time as the boatmen on Loch Monar and undertook in that capacity to row materials or provisions as required up and down the six and a half miles for the sum of one and six pence per ton. Some time later, Farquhar MacPhail became the Gamekeeper at Strathmore but his life was full of tragedy, as the following deeply

moving inscription in Struy burying ground indicates:"Erected by
Farquhar MacPhail, Gamekeeper, Strathmore. To the memory of his
beloved wife Mary MacLennan who died at Strathmore, Nov. 1st 1886
aged 57 years, and their children, Hector died 25th April 1889 aged 7
years, Annie died 14th May 1888 aged 9 years, George died 18th Feb.
1898 aged 22 years. The two younger children died from poisoned
water but the boy of 22 years of age tragically committed suicide
beside the swing bridge, a wire and timber construction which
spanned the river about half a mile west of Strathmore. I could be
correct in surmising that MacPhail's wife might be a sister of Eachan
Mhor (Big Hector) who lived at the little croft to the east of
Strathmore and whom I have already mentioned. By every account
MacPhail was a fine Highlander and bore his grief with dignity. In
these days of health care at every hand it is perhaps difficult to
appreciate the depth of loss that sickness and death would effect in a
lonely homestead. The hills bespeak a healing power and closeness to
unseen help now lost in the security of modern welfare. Those who
befriend the silent peaks in a lifetime's work know well their
embracing strength.

At a more recent date old Kenny's brother Farquhar MacKay,
known as Fachie, worked as ghillie boatman at Pait. His days in this
demanding job came after the First World War. Often the fashion, Pait
shootings happened to be let one year to an English Gentleman, who
took with him a chauffeur of the same nationality. The ghillie
boatman was instructed to give the newcomer some lessons in
handling the motor boats in order that his chauffeuring expertise
should be extended to journeying down for the mails when stalking
duties occupied Fachie and the other boys. Duly, the boatman, a party
of ghillies and the disdainful chauffeur set off one afternoon to collect
the postal delivery which in those days arrived at Monar from Struy
by way of George Tait with his horse and trap. Need it surprise you to
know the Pait hill men were cordially invited to their bothy by the
Monar ghillies who, idle for the day due to mist, welcomed a
diversion. The English chauffeur conscious of his superior position in
relation to 'The Gentleman' and possessing all the pomp and
circumstance common to his race disdainfully refused their welcome.
Postie Tait came and went. By and by it became clear to the impatient
Englishman sitting disconsolately at the pier that a typical Highland
carousel of a generally undignified nature was rapidly gaining control
of the day. Being at once excitable and disapproving, with Nelsonic
resolve he jumped into the launch and made off up the loch at speed.

Doubtless intent on laying before his 'Gentleman' the whole unrefined and disgusting deviation from duties implicit in the behaviour of the revelling ghillies.

Now time is of little relevance in the Highlands and a certain expendable quantity had elapsed before the merry makers emerged to discover their supposed comrade's desertion. Enraged they gave chase in the Monar motor launch. Sighting the sandbank at the west end of the loch they quickly discerned the zealous chauffeur clearly possessed more knowledge of motorised transport for he had run the Pait boat hard aground on the sandy spit. Worse, the boat remained in gear at full speed ahead futilely churning sand. The Monar launch pulled alongside crewed by what must have appeared to the southeron as a piratic horde in ugly humour. Fachie being a strong agile man leapt aboard the stranded vessel and seizing the now trembling runaway, threw him over the side. The chauffeur came up gasping for breath, only to be thrust under again by Fachie roaring in Gaelic to the effect that the Englishman's presence was no longer required in Monar. Thinking his life in danger the unfortunate man broke free, got ashore and ran helter skelter to the Lodge crying to the shooting tenant that an attempt on his life had been perpetrated. Fachie lay low. The English gentleman immediately complained to the Proprietor, at that time a Colonel Haig. Down came a summons to the bothy.

The Colonel wished to see Fachie privately. Subdued ghillies felt in no doubt that in addition to a severe dressing down, MacKay would be dismissed on the spot. Prepared to take his medicine, the boatman stepped smartly into the proprietor's study in the Lodge. "So Fachie" frowned the Colonel as he swung round in his chair "you tried to drown the Englishman did you?", and gave him two pounds.

A NIGHT SAIL

All the dogs appeared to enjoy a jaunt in the boats. If I were going across to Pait for any reason then I had eager companions. Nancy and Sheila realising that my step was making down to the boathouse capered ahead, tails up and eyes bright. As I came down to the pier they would be sitting in one of the boats. Should it happen that I got into a different row boat, or perhaps the launch, no word was needed. With the alacrity of a midshipman joining his first vessel they'd skip aboard the one about to leave. Once under way the dogs sat on a thwart balancing to the motion, heads up sniffing the breeze or watching the water for a splash from the oars all in the manner of grand ladies generally enjoying the freshness of the day on the water.

All was not necessarily peace and harmony should many dogs be obliged to sail together. Our dogs hated the MacKays' canine contingent to trespass on one of the Strathmore boats. Fights would break out below decks, often ending with the aggressors being flung overboard as a measure of discipline. The venomous look of reproach in the eyes of a vanquished dog swimming for the shore gave us a degree of malicious delight. Often the cocky dry dogs leapt for the pier as the boat bumped in and raced around to intercept their bedraggled foe wading up the stones further down the shore. Dogs always believe in kicking a man when he's down it would seem.

Wily and purposeful to the last, Bob the beardie played his farewell trick on us in the sure form of a litter of ten of the bonniest, cuddliest pups that we were ever to have born. Nancy produced them in a straw bed at the back of the old wooden dogshed one night some four months after her clandestine lover had been despatched to exercise his talents in the fertile pastures of the Black Isle. I had thought from the swelling and filling of her nipples the puppies' birth was imminent and accordingly that morning after milking the house cow, looked over the double doors of her shed. Dimly at the far corner I saw the bitch lying on her side. Into her teats were pushing ten little black and white bodies, back legs stretched out and tiny stumpy tails waving. Squeaks, snufflings and sucking noises told of a competitive breakfast. Nancy looked up with happy eyes and gave a single wag of her tail. I closed the door and took the news down to the children. Their excitement knew no bounds as they scampered up the track past the byre to the kennel door and stood peering through the cracks, having been warned on no account to go inside until the puppies were a day or two older. Soon they grew into fat, little dumplings running about on strong stubby legs at their shed door and amongst the surrounding rashes. At this stage they were the best imaginable pets. Alison and Hector stayed hours with them inventing various games. Races that went wrong, hide and seek with victims that wouldn't remain hidden, or jump off the wheelbarrow. Lovingly squeezed puppies looked wildly about for help, showing the whites of their eyes in alarm as they were trundled round in the old pram squirming about with a doll's jersey pulled over their heads. Nancy kept careful watch but a bitch of splendid temperament she never once indicated any snappiness with the tormentors of her litter.

We selected a strong, well marked dog pup which had a black roof to his mouth and called him Shep. He grew out to be a powerful thickset chap with a long rough coat. Handsomely black and white his crowning glory floated bannerlike in the form of a large wide bushy tail, white haired at the tip. At the age of twelve months shepherding duties began to fall his way and I found he had a mixture of attributes from both his parents. Ultimately when trained he would bark to command, drive, gather and, always standing upright, could move a bulk of sheep along with great gusto if they began to flag. Bob, would have smiled to himself as early in his son's career we began to appreciate that Shep inherited his sire's penchant for a fight. Like his old boy, the moment Shep set eyes upon any strange dog he would adopt a beligerant posture. Without doubt, apart from bared teeth, his most adversative feature became the huge broad bushy tail which he raised straight up and held stiffly erect as he advanced deliberately on some luckless cur.

Poor Shep, he was to have his fighting prowess reduced in a most painful and humiliating way. Apart from rowing across, pier to pier, at our end of the loch, not often did we find ourselves out with the boats at night. However it occasionally happened and one October evening I returned late from a sheep sale in Inverness. The journey had been delayed, pleasantly I might tell, by stops for tea, first at Braulin with John Venters the Lovat Estate shepherd, and then, much later, with Allan Fleming, the Monar keeper and his family. Towards two o'clock after bidding Allan and Gracie goodnight, I walked the rough half mile track winding down to Monar pier from the keeper's house. Few stars were to be seen as I stepped out on a fresh lung-filling night. The rapidly moving clouds told me, as I felt for the weather, it might be blowing hard further west the loch. Monar pier and its waters situated at the sheltered lower reaches of the loch encircled by high peaks, could be deceptively calm. This was by no means a guarantee of the conditions prevailing once boats cleared the cliff overhung narrows connecting the Monar basin to the main extent of Loch Monar. I had taken Shep down to Monar with me that morning leaving him tied in the old garage twenty yards up from the pier. As I loosed him, the wood chips scattered about the floor indicated how he had been occupied that day, any confinement he found most irksome. Down to the pier he dashed, out onto the boat, twice round the deck to check for any canine stowaways before together we set off for home.

Moving safely on the loch at night required the crude navigational skill of judging one's position from the dark outline of the hills against the cavernous sky. A close familiarity with miles of surrounding ground made this method relatively simple and safe. Navigating on really black or misty nights tended to be neither simple nor safe and might be better described as random reckoning. Passing through the narrows that particular night, the cliff towering above me ten yards on my right hand echoed to the chug chug of the *Spray's* throttled down engine. I felt the darkness, I could touch some new element, it hung in the dank air, pressing close, etherial, yet within human communication. I shivered and gripped the tiller. The boat slipped smoothly through the still channel. Water circulated from the cooling system to splash rhythmically out astern with each cough of the engine. Steam curled upwards mingling with the exhaust. Eyes and senses were tautly alert. I steered carefully through this dog-legged stretch relieved to clear the steep rocks at its west end before swinging the tiller hard over to point our bows to the north, where the outline of Craig na h-Iolaire now reared distinctively against a brightening backcloth of late autumn stars.

Moving out into the wider waters, the boat began to take on motion which indicated a strong swell was running, what we described as a 'good heave on the loch'. An increasingly luminous sky impatiently shed its scudding cloud cover as the wind freshened to a stiff blow from the west. Ahead lay the hills of Strathmore, sharp edged and darkest blue black, glinting stars starting to dance on their peaks, and the huge rounded bulk of Maoile Lunndaidh lay to the north west like some sleeping form in a distant world of forgotten time. Without warning, to the south of me, breaks of driving cloud released the moon. I turned to wonder at the closeness of its yellow form, full of the season's harvest as only an October moon can be. The golden orb dwarfed with its nearness the curving shoulder of Sgurr na Lapaich, now an ink black silhouette etched out behind me. Ahead the waves lost their sullen menace, each tumbling top pranced to me with living light and the breakers on the far shore below the gentle Maoile Choill-Mhias, showed white as a lover's teeth. For only once is it given to man to lie in the arms of the eternal forces which give all matter form.

The *Spray* drove into this primitive world. She felt as alive to me that night as would have been the pine forests which gave her timbers birth. Fine trees, they knew such same stark splendours from some lofty pristine crag before giving all to shape the form of my lively boat. We sailed on midst a great universe in which I felt borne that night through the wild embrace of the elements that raced about me. The wind on my cheek flung the waters from each wave, dashed by the bow to wet my face as we sailed to a horizon that led over the nearing peaks of home to a friendly space which knew all things been and gone, and all things yet to come.

Shep, it might seem, shared no such fanciful conjectures. To my concern he jumped out of the hold to run about on the narrow swaying deck. "Get down boy" I yelled at him. The spell of the night was broken. With an indifference, typical of his father, Shep ignored me, held his head up into the wind and continued to perambulate the deck. Thoroughly annoyed and not a little afraid that Shep, with his Flag Admiral swagger might see his pride end over the side, I jumped smartly out of the engine cockpit, caught the sailor laddie by the scruff of his neck and took him beside me into the engine hold. "Sit there now and settle yourself you stupid dog". Suitably dousing his hubristical arrogance, I added a sharp kick to emphasize the instruction. Shep was much subdued and probably frightened due to the clanking of the engine at such close quarters. I dimly discerned that tail down, he slunk to and fro in the confined space behind my legs.

Now well out towards the widest waters of the loch we bucketed into good sized rollers. Ignoring the demoted hound I kept a weather eye on the biggest white tops, taking them at three quarters bow and rolling the boat over the high crests rather than driving straight through them. Indeed I was revelling in the challenge when to my horror the engine note changed and it began to labour excessively. Slowly the running rate fell. Curse, a rope fouling the propeller flashed my instant reaction. In these night conditions this spelt considerable danger. Using my foot, I spun the clutch wheel to put the engine out of gear. Freed of load the engine leapt back to life.

Only then did I recognise the howls and wails of anguish that issued in high piteous notes from under the engine room seat. Grabbing my torch, the beam revealed poor Shep hard down against the stern post, biting frantically at the sides of the boat. Momentarily I was puzzled, but as the propeller ceased to spin Shep heaved and struggled pulling his hind end out of the bilges. I saw then the cause of our near power failure. His magnificent tail had caught and wrapped around a small exposed section of the propeller shaft with a sufficient grip to virtually stall the powerful engine. What a sorry chap he looked, turning to whine and lick himself. I lifted the poor dog over to the hold and hurriedly regained control of the launch. We journeyed home more quietly as the wind began to ease. Fortunately Shep's hind legs were not damaged, but for months afterwards his tail hung like a damp rag trailing on the ground. The rampant fighting banner never recovered and remained only a token of its former waving glory. In due course, after what could be described as an off season, Shep's fighting spirit revived, though barely to its previous invincible fury.

The general care and maintenance of the boats had always to be a priority that occupied a foremost role in our daily lives. Each morning noting the strength and direction of the wind, the state of the loch, its height at the piers, whether heavy rain might ease or cause a flood - all factors affecting the boats - became second nature. Ice during the wintertime often gave cause for concern. The head waters of Monar between Strathmore and Pait, being shallow, froze each year. Most surprising was the rate at which it could freeze. Going down for the mails during the four months from early December often involved dealing with an ice problem. Leaving for Monar about middle day, the Postie would be met, assuming there were no snow difficulties for him travelling up the glen, at the shed above the pier around 2 to 2.30 p.m. A wee pause to gather some local news was a must. Cossor the Postie would deliberately light his pipe and though his extreme deafness made conversation difficult he was always cheerful. Whatever the conditions he obligingly brought up passengers or extra parcels without complaint.

Postal delivery from November through to March depended upon the weather. If a winter's day had been clearing away to thin brittle sunshine then by 3 o'clock you would feel the frost settling like a cold mantle. It was time to head the *'Spray'* home from Monar. Watchfully into the narrows, still and black, for this stretch might already have a skin of ice. Sure enough at the first bend of the dog's leg channel a chinking crackle would be heard up front. Immediately cutting the boat engine to its slowest revs we edged forward through ice panned water leaving tinkling slivers of ice bobbing about to mark our track. Usually, with two on board, one person went to the bow and broke a passage using hefty swings of the boathook. By this time, the sloping fore deck of the boat polished itself a film of ice – thoughts of the coldness of the water ensured a masterly sense of balance. Out into the loch and clear of the narrows we'd open the throttle to full revs and make west with all speed. Calm but apprehensive we watched our wake spread its 'V' to a great width in the thickening waters, sometimes reaching from shore to shore. The chug chug of hurried progress carried far on crystal air and on such evenings reached miles ahead of its source. Looking astern to the hard conical shape of Beinn na Muice, a bare inhospitable mountain thrust onto our eastern horizon, the sky gradually turned deep mauve, sometimes shading out to red before a peculiar translucent lightness of blue higher in the sky told us a keen night's freeze must be expected.

The ice at the west end of the loch could now have formed to a thickness which might prevent us easily getting to the piers. We invariably carried provisions and to save a carry we generally tried to get ashore at the jetties. Nothing for it but another balancing act at the bow and break our way in. Any attempt to force a passage cut the timbers of the hull in a very short time. If we considered the frost might hold then it was straight back out with the boat and round to the 'new boathouse'. This alternative facility had been built on the Pait

side, east of the headwaters, in about 1920 as a pier and slipway to serve
on such occasions and also to provide cover for painting or general
maintainance. The short winter days left no time for deliberations.
The *Spray* crackled her way rapidly out through the channel just
broken, across the corran, and down to a safe berth for the night. The
open channel closed swiftly behind her, a lattice of broken fragments.

The importance of a continuous observation of wind and weather
pointedly came to me one night in a most significant way. It had been a
day of gale and flood in early November. I romped down to collect the
mails, westerly force eight speeding the *Spray* down to Monar.
Running before wind and wave gave the exhilaration of the *Spray*
gathering speed as a big roller bore her along like a cork and passed
foaming along her sides. I felt some reservations about getting home
however, as steep breakers reared astern towering above me before up
rose the buoyant boat, lifting me high above the plunging bow.
Thinking the wind eased back, I set course for Strathmore with my
letters and groceries, and after a struggle made the home pier. The
wind rather than decreasing was in fact veering to the north. From its
new quarter it continued to blow just as hard. The trick for landing
and tying up when alone in gale and flood required running the bow
against the pier, leaving the engine running slowly ahead then diving
to the bow, grabbing a rope and leaping into the water, hoping to land
on a solid structure. . . On this inclement night the lower end of the
jetty lay submerged waist deep due to the flood. Add darkness and a
sleety gale and securing the *Spray* became a difficult task. I fumbled
with numb hands in four feet of icy water, virtually diving to find the
pier's mooring chain. Mooring ropes secured stem to stern I left the
boat safe for the night and ran dripping home for a hot drink and
change.

A cosy evening passed with a book before the fire, feet on the
stove. The wind away from bullying our west gable came soughing at
the little back window - away to the north I thought through the pages
and read on. A last yawn and so to bed, in our lifestyle seldom before
one or two a.m. The wind seemed to drop away, the rain had stopped
hours ago. I knew the flood would be going down perhaps as quickly
as it rose. Maybe it was a rattle on the slates, I don't know, but I
wakened sometime later fully alert, sensing that the wind had swung
round to the north east. Some sixth sense told me the *Spray* should be
checked, though there could be no reason but to believe she lay safe as I
had moored her carefully. Pulling on trousers and jersey I ran down to
the pier. The sky had brightened, a face biting wind held strong. In

consternation I saw as the moon burst through racing clouds the green hull of the *Spray* heeled over, perched high on the top of the pier. Had a malevolent spirit, intent on inflicting reprisals for my savage cursing of its sacred powers some hours previously, arranged a shift of wind which swung the boat across the flooded pier just as the loch began to drop? I made a note to adopt a more reverential tone of communication with the elements, as a vision of the boat lying stranded like a whale waiting for the next high tide flitted through my mind. Uncomfortable thoughts; a salvaging tide might only flow when nature chose her next major flood – six months away, who could know?

The *Spray* weighed several tons. A critical situation erred on the side of understatement. Waves still broke along the pier. I plunged again into the icy water, fumbling for mooring ropes. The moon came and went. Checking a curse I finally got the ropes free. I heaved and strained to no avail; no time to run and call Betty. Repentance worked, a bigger wave swept in, I strained on the bending pole, at last a move from the dead boat. With redoubled vigour I strove, levering with the pole as each wave passed under the hull. Foot by foot, the *Spray* slid off the pier and bobbed into deep water. I joined in her relief. Soaked and not a little cold for the second time in a few hours I secured her much lower down the pier and ran home. Back to bed, warm my feet and tell a turn of fate.

The children, knowing no other, regarded the dependence on moving anywhere by water as a fact of life, though they were not allowed to go down to the loch side unless one of us was with them. Such discipline was applied not without good reason. After the MacKays left, the boats became my responsibility. Time for maintenance fitted into the day's work, adding interest and experience, One morning I set off for the boathouse to put in an hour or two's work on servicing the *Spray's* Kelvin engine. The day bright but chilly attracted Alison and Hector, who with gloves and duffle coats on, skipped down to the boathouse at my heel. It might seem there could be little to occupy their interest but once down on the pier the children made their own games. Climbing on to a big rock and jumping down, in and out of the rowing boats, generally enjoying life with an inventiveness which youngsters not overwhelmed with toys soon adopt. I took the cover off the *Spray's* engine cockpit and clambered inside to busy myself enthusiastically, as amateurs are wont to do, with the removal and cleaning of the sparking plugs. By and by I heard Alison's feet slipping down the deck, a face peeped in.

Promise of a fine day. Shepherd's house, Strathmore

Looking west, Loch Monar on a typically overcast day

The cluster of Pait across the head of Monar with its sandbank, boathouses, and the Spray at anchor

The high top of Strathmore ground Bidean an Eoin Deirg from the ruins of the deer watcher's house at Cosac on the south shore of Loch Monar

Erchless Castle, fifteenth century seat of The Chisholm

Sgurr na Lapaich

Ardchuilk, once the home of John Ross (see Pait Blend)

Braulin Lodge

The woods of Strathfarrar in springtime

The Monar Falls on River Farrar below a gorge gateway to the vast area at the west end of Strathfarrar

One of Scotland's few remaining stands of Caledonian pine in Strathfarrar

Loch Beannacharan with the long ridge of the Lapaich far to the west

This single pine is visible from many miles on its sentinel rock in Strathfarrar

Author in the sheep fank at Strathmore

The hoggs come home from wintering

Alison and Bob the Rascal
Pait with beyond the fine long ridge of Riabhachan mirrored in the calm of a spring morning.

Coire Shaille, flanked by the cliffs of Bidean an Eoin Deirg

The great Corrie of Toll a Chaorachain.
The peak on the left is Sgurr na Conbhaire, right, Sgurr Choinnich

Achnashellach Forest viewed from Sgurr Choinnich

The Strathmore croft and across the river flats to the shoulder of Meall Mhor

The shepherd and family

A spring morning with the midden in heaps on Strathmore croft

Hinds for the salt barrel. L. to R. author, Kenny and Iain Mackay

An easy drag on the snow

At the Strathmore peats

The hidden falls of
Allt Toll a Chaorachain

Minutes old, a cheviot ewe licks her new born lamb

Betty and Alison with Dandy carrying a lambing shelter to the glen

Spring morning, ewes and lambs away to the hill
In April a load of sheep feeding goes west to Strathmore glen for the lambing ewes

Hungry spring. The ewes, heavy in lamb, feeding on the Strathmore river flats

Storm clouds threaten Maoile Lunndaidh

The rocky bed of Allt Mhuillich with Creag na Gaoith above

The Spray *anchored at the head of Loch Monar*

The Spray *with Iain at the tiller, a 'rogue jeep' arrives*

Mischief at a month old, Nancy and Bob's puppies

The Spray and An Gead lie at Strathmore pier

Reluctant pets, with Alison and Hector
The Mackay boys. Kenny holding a deer calf and Iain with Bess. The Glen Orrin cattle drive.

Glen Orrin head with Loch na Caoidhe. The tracks marking the hillside in the left
foreground are ample evidence this pass has been used as a drove road for generations.

Glen Dhomhain

Forcing the cattle to swim the corran at the head of Loch Monar

Keeping their feet on the bottom

The leaking old coble

The milk ewe dipping at Strathmore fanks. L to R Iain Mackay, Big Bob Cameron,
Old Kenny Mackay rolling wool, author, Young Kenny Mackay, Sandy Thomson,
Alison

At the hay, taking loads off the fence to the barn

'Coilacks' of hay await a brisk day

The famous whisky smuggler Hamish Dhu of Pait

The thatched house of Hamish Dhu Macrae and the causeway built to civilise his island retreat

A panorama from Creag na Gaoidh: above the Gead Lochs Beinn Bheag, An Cruachan, Aonach Buidhe and Feachag

The peak of Bidean an Eoin Deirg high above Loch Mhuilich

From a small lochan on the slopes of Beinn Tharsuinn the shapely peak of Bidean a Choire Sheasgaich

*Ready for the pony two fine
stags lie gralloched high on
Maoile Chioll'Mas*

Dandy loaded with hinds

On An Riabhachan, stalker Old Kenny Mackay with the last stag he got before his retiral from Pait Forest

Pait Lodge

Nancy and Sheila. Loch Monar frozen from end to end

Moving sheep on frozen Loch Monar

...ain Mackay and Dandy out on the ice at Strathmore boathouse

...unset on the high Strathmore hills

The end of an era

Cutting and burning at Monar to make way for Hydro power

Ready for the dynamite. The Strathmore home

The horse shoe Monar dam

Loch Monar today

Isolation Shepherd

"Daddy", she said quietly, as though come to tell tales of a friend, "Hector's in the water with his new duffle coat on". I leapt up on to the pier and with sickening dread saw Hector floating face down some yards out from a rowing boat. In a flash I grabbed him out, his eyes were closed. He didn't seem to be breathing. I laid the little chap down on the pier and applied artificial respiration. Water came out of his mouth and then a sound. He started to cry and cough, all seemed well. Taking him in my arms I carried him back to the house. Once he recovered we got fanciful stories about the fish he had seen under the water.

Navigating on Monar was relatively straightforward in the sense that water levels remained in the hands of the weather and only during exceptional floods did they alter a great deal. Boatmen knew where the hazards lay and steered accordingly. Not so on Loch Mullardoch, the huge sheet of water in neighbouring Glen Cannich. Here due to flooding for Hydro-electric production, two natural lochs had been joined together. However the area between them remained shallow, hags and hillocks made navigation problematic. Our friends and neighbours in this glen were Jock and Dolian, stalker and shepherd-boatman respectively, the latter's name being a combination of Donald and Iain. Both were characters in their own rights, but Dolian from the west, laid particular claim to the humour and flavour of natives from those parts. During the stalking season they would be as busy as ourselves taking care of the shooting parties. Entrusted with a singularly influential group of sportsman west up Loch Mullardoch one September morning, Dolian handled the controls of their speedy launch which he drove along, throttle wide. In the passage between the two lochs all looked water for miles around. Whilst guests exclaimed admiringly at the scenic splendour, the Proprietor felt more concerned with navigational matters. "Are you sure it's deep water here Donald" he queried sharply of Dolian. "Oh indeed Sur, it's as deep as the Pacific." The confident words barely left Dolian's mouth in a charming Highland drawl, when with a sickening crunch the boat stopped dead. Everybody including the scenically spellbound notables were flung to their knees. Two of the more engrossed narrowly missed going overboard. "Oh my God" and similar utterances indicating stiff upper lip British aplomb thinly disguised a moment of terror. The launch had run hard onto the top of a hidden peathag. Full astern made a flurry of water but no other visible effect. "Now Donald this certainly will not do" remarked the Proprietor in a tone which gave Dolian a hint of displeasure. The boatman, immediately perceiving this to be a situation not remediable by west coast charm, clambered up to the bow and without a word lowered himself over the side up to his chest in the cool but tranquil waters. Careful to keep his footing, he put his shoulder to the stem. Jock the stalker, equally subdued, though generally a match for any Gentleman's irritation, ran the engine in reverse. Dolian heaved manfully at the bow. The shooters peered dubiously over the side, one of the party making feeble prods with an oar. Without warning, and matching the suddenness with which she struck, the boat shot off the bank at full astern down the loch. Dolian was left standing on a slippery peat hag surrounded by miles of water. Waving frantically he stood out there up to the armpits, in gently lapping Loch Mullardoch, informing the rapidly reversing launch party with something less than his usual plausibility that he couldn't swim. As Jock recalled "he stood there dabbling about for all the world like a heron fishing."

No Highlander has any doubt about the existence of 'the second sight' or indeed simple prognostication. Many old stories told around an evening at the fire would hinge on such happenings. The last Christmas we were in Strathmore I had a most unusual manifestation of this power. The old MacKays had left that previous summer to retire, and the boys to take up new employment. We became a family alone at the west end of the loch. On the afternoon of Christmas Eve I set off with the *Spray* to Monar to collect the mails. The weather held fine though the clouds suggested to me that a change could be on the way.

No therapy could be more relaxing than a forty minute sail down the loch in good weather conditions. The hills for those sharp of eye were always full of interest. Perhaps it might be noting the movement of deer on Maoile Choill-Mhias or spying at sheep down about the halfway house in case they were belonging to Strathmore and had strayed too far east. An interlude at once peaceful and thought provoking centred in an existence upon which the scrambling world bore little effect. The chugging cough of the good old *Spray's* engine contrasted against the constant song of water under her bow.

The Postie came late to Monar that afternoon and I walked the half mile up to Allan the keeper's house, to save the post car from the rough track down to Monar pier. Allan's wife Gracie always made a cheerie welcome, and I gladly stayed for my supper as it was fully 7 o'clock before Cossor the Postie arrived. We sat talking until a rattle of hail on the house windows warned me the weather blew for a change. I did not delay. Heaving a bag full of Christmas mails and parcels on my back I ran down the track through gusting sleet. The *Spray* lay waiting, a dark shadow against a skiff of snow which began to gather on the planks and posts of Monar pier. A couple of smart pulls on the starting handle and the *Spray's* heavy engine rumbled out into the night. I slipped the stiffening mooring ropes, and we eased astern from the pier, water churned white under the rudder in the faint light.

It was Christmas Eve. I set off for the isolation of Strathmore, happy and carefree at the prospect of playing Santa Claus to eager faces. Clearing the narrows I opened the throttle, for the night seemed not so bad as I had feared. Though there blew northerly sleet showers, the wind relaxed between each blast and white-edged hilly outlines stood clear enough against the drapes of grey cloud for easy navigation. A rolling swell took the bow and gave motion, but I decided all should be safe and accordingly set the tiller on to self steering. This, I should say, meant nothing more elaborate than tying it with a piece of string. Once the boat was set on the desired course, I duly sighted the bow towards the bold outlined summit of Maoile Lunndaidh, and tightened the string to hold a course for me. The main hazard at this point on the loch was a long rocky promontory which ran out some considerable distance from the south shore. Its black wave-beaten sides dropped sheer into deep water, whilst right out on the exposed point stood a single Scots Pine of girth fit to match any storm.

Allowing for this natural peril in setting the tiller, I dodged below and crouched beside the engine, where, with the aid of a torch, I began to read the many Christmas cards. Many wished us well, wondering, no doubt in the manner of concern for lighthouse keepers, just how we fared. The *Spray* cheerfully sailed on without my guidance. Half way through the welcome pile a strange sensation drew my thoughts from the greeting cards My mind was detached from the boat, the noise and the dark wind-rustled night. As if in a dream I saw myself standing in a study, a large room, comfortably appointed with oldish furniture, it contained books and paintings. I spoke in a concerned manner to Sir John, my employer who sat in a swivel chair looking most intently at me with his bright keen eyes. In some form of explanation I heard myself saying "I'm terribly sorry Sir, the boat is a complete loss. I can offer no excuse. I was down beside the engine reading Christmas cards." He looked up at me without apparent annoyance but didn't speak. I saw him and the room with all the unnerving reality of actual presence. It passed in moments.

For a second I felt struck by horror as though in the grip of some awful tragedy. I stood up swiftly, straining eyes into the blackness. Instantly the true horror flooded through me. Just yards ahead towered the promontory. White tongued waves swirled with evil menace along cracked life-hungry rocks. I spun into reverse, held my breath. Slow sickening seconds ticked through my stomach. Almost too slowly the *Spray* lost way, there was the slightest bump before she mercifully pulled astern. The black cliff looked sheer, ugly and animated with the evil of a death cheated spirit. I felt quite shaken and did not trust the string trick again that night. Coming home I tumbled out the story of my escape to Betty, who became annoyed at my risk taking, and not a little uneasy at the strange premonition. For certain the boat would have been lost and perhaps me along with it. She could have waited alone with two small children in a wide uninhabited world, for the dawn of an empty Christmas Day.

Many years later, possibly fifteen or twenty, Sir John becoming old, asked if I would go to see him. This I readily did, for fewer men I have held in higher regard. I had not been to Fairburn House before – a large Victorian mansion. A servant came to the door and I was shown in to see the old Gentleman. To my astonishment I recognised the room. I was back in the dream, Sir John seated in the same chair looked up, his eyes bright and welcoming, just as I had seen him all those years before in a vision which that Christmas Eve perhaps saved my life.

DROVING HOME

Summer days found Strathmore at its most idyllic. The fullsome green of the hills were patterned in vagrant shadow, cast by aimless clouds that sailed through blue skyed days, when all life relaxed and strolling time took its step from the sun. Though much work needed to be done, no sense of urgency pressed on the weeks that slid easily by. Although we were all employees nobody planned our days. Few people arrived without we knew of their coming, and this relative security from intrusion encouraged a lifestyle which in many respects resembled that of the days of a now forgotten era. Work did not seem arduous or boring. Often myself and the MacKays would share tasks in the old communal way with much fun and banter. Peats cut in May had to be turned and stacked. The winter's accumulated midden of dung at the byre I barrowed down to the croft and spread over the acre of grass, which later in the season became hay crop. Repair jobs of a variety of kinds, fencing, the pier or the paths, all were tackled according to the day and one's own inclinations.

Early in June there came word from the home Estate that the hill cattle which grazed annually in our glen would be moving out from their winter quarters on a certain day. Could we come so far along the route to meet the Fairburn boys and help with the drive? In our customary fashion the MacKay boys and myself decided to turn work into play and planned an expedition accordingly. The route the cattle trek took from the home farm at Fairburn led up Glen Orrin past Corrie Hallie Lodge, and thence to the Bealach over into the East Monar ground at Corrie Dhomhain. This long narrow steep sided glen bought them down to the Monar lochside for the final trek west to Strathmore.

Dandy the pony was commissioned and fitted out for the trip with a deer saddle, hung either side with roomy paniers filled with enough requirements for a modest expedition to the Himalayas. We left after breakfast on a breezy morning of light rain which promised to clear. Three lank lean figures striding out down the lochside in high good humour. Dandy also stepping out in fine style seemed glad to be on the move, enjoying himself in spite of our teasing. We didn't normally have Dandy shod unless he worked hard at the deer carrying during the stalking season. What a sound hardy chap. He always worked with a great will, though not above a trick or two, and who could blame him. The day brightened as we climbed out of Corrie Dhomhain and made the steep descent into the head of Glen Orrin. The fresh clean wind held a taste of the snows which still lay on the highest ranges stretching away to the north. Our spirits soared with a touch of adventure. At the first lochan we stopped for lunch. Our first precaution taken with the mischievous Dandy was to tether him whilst we ate. Not too short, so that he could lunch along with us on the crisp short grass at the lochan edge, and securely enough to ensure he wouldn't, if feeling homesick, depart suddenly westwards at a smart clip. I say securely, for the cute Dandy was not above playing with a tether until he got it free, then with a schoolboy slyness, the little brown delinquent would sneak off home always keeping tantilisingly ahead of a pursuing ghillie. It was a fine bright day, so out came the fishing rods. We tried the lochan but perhaps the sun was too strong, for we didn't catch our supper.

Abandoning the fishing, we reloaded Dandy and trekked on, following the river to where the glen broadened out into grazing flats beside a bonnie lochan. In these pleasant surroundings had been built Corrie Hallie Lodge. It lay empty and disused, we promptly decided to commandeer it for the night and set about unloading and installing our kit. Whilst tethering Dandy out on the flat below the Lodge I spotted a red deer calf. It lay flat to the ground with barely a breath of movement except for the large puzzled eyes which swivelled to watch me as I quietly approached. Each year we came upon deer calves on our travels about the hills; this was the first calf I had found that season. How well a russet coat blended with last year's lank dead herbage, flecked white spots did not detract but enhanced its camouflage. I looked down, not a sign of movement did it make, ears flat along the neck, barely breathing. As I lifted the tiny delicate object, all eyes and ears, it played dead, just a slack limp bundle until after a moment or two, without warning, it let out a piercing high pitched squeal and struggled out of my arms. The hind, which I hadn't noticed, had been watching from the slopes above the Lodge and now without fear came running down towards us. Bravely she came within fifty yards before hesitating. Mothercare versus fear of humans, the hind stood head up, watching. Her calf, in the nature of the very young clung to the nearest object, friend or foe, and had not run from me. As I began to walk up to the Lodge the baby creature followed, gangling along on slender awkward legs which didn't yet quite do what the calf expected. Kenny seeing the dilemma from the Lodge window came out. "It's got you labelled as Mother now", he called over to me. "Never mind the comedy, come over and see if we can get rid of it" I ventured. Between us we caught the calf again and carried it up the hill a little way. The hind, unable to face our activity, had vanished, but we guessed or hoped she would not be too far away. After several tries at forcing long thin legs into a folded position to make it lie down, we eventually got away without the calf jumping up and following. During the evening we watched the hillside on and off but the hind did not return. However by the morning the calf had gone, we trusted with the correct mother.

"How about giving Sandy a fright" Iain suggested, for at long last, in the early evening, we heard and saw the large herd ambling up the glen towards us at about two miles distant. A spy with the telescopes told us there were two men and a pony following the slowly moving cows and calves. Iain's idea caught on. Straightaway we gathered up armfuls of dead bracken and piled it inside the lodge porch and round at the rear windows. Judging the oncoming drovers to be half a mile away, sufficiently close for the plot to work well, we lit the bracken, some of which was rather damp material. The wind direction helped, in minutes billows of grey smoke rolled out of the old building's door and windows. "Quick" directed the chief prankster, "pretend to be running to the burn for water". Kenny and I dashed up and down to the water as obviously as we could without actually having pails. Iain sat with the glass. "They've seen it" he called to us as we acted the part of the local fire brigade. "Sandy's stopped, they seem to be talking. Now Sandy's running and waving". The commentary kept us laughing as we pretended to quell the blaze. By now Sandy the Fairburn shepherd was not far away, but the fire and smoke died down. Sandy stopped in doubt, much out of breath we hoped and suspected; not sufficiently puffed however to prevent him shouting something of an undignified nature in our direction. Our would-be saviours turned and went back to droving.

The cattle lumbered up to the flats and started to graze, seeming to realise they were to rest there for the night. The two tired Fairburn boys, Sandy Thomson and Willie Pirie strolled in, their pony, like ours, kitted out with loaded panniers. At once Sandy started, "You fools, where do you think we could stay tonight if the Lodge burnt down?" To our delight it transpired the ruse had worked; the drovers were convinced we had set the Lodge ablaze - not that they gave a rap for the building, but definitely preferred a night under a roof to one gazing at the stars. In pseudo wild west style we kindled a cooking fire and made a chuckwagon meal - salt venison, potatoes, milk and rice pudding, plain fare but wholesome. A good supper, a sharp evening of long summer light and, to the Fairburn boys' surprise, we took a football from our panniers. No persuasion needed, we drove some loitering cows from a level area and the glen echoed to a good hard game. The cattle looked on with cudding half interest, some feeding their calves before settling for the night. Dandy made friends with Shean the Fairburn pony. A half moon peeped over an eastern ridge as tired and happy, we rolled into blankets on the old Lodge floor.

Next morning, after only a few hours sleep and a quick breakfast, we got the cows to their feet ready for the main part of the trek. Checking round the herd, Willie the cattleman found one of the cows had calved overnight. A cross Aberdeen-Angus calf, small but lively, Willie told us he doubted if it would make the journey but "like the rest of us" he said "it can take its chance." Kenny and I saddled the ponies. Shean seemed to catch the excitement of moving on and skitted about as we tugged at belly bands and tail straps. Whoops and shouts, waves and sticks, agitated cows called up their calves, bawling and milling a little until some of the lead animals picked up the path away from the river flats to head west up the long glen. We were in fact travelling an old drove road, the baring of the stones over a broad area told of many hooves passing that way before us. A pleasant day of cloud floating south, a west wind warm and growthy, pulling the cattle back to sweet hill pastures that surely they recalled. Moving along with willing step the lead cows picked out the winding route. Men and ponies followed a cacophony of calls, anxious staccato calves against a background of throaty answering mothers. The herd swung along to a clip and rattle of stepping hooves, swinging tails and rolling backs strung out ahead of us   The air hung with the smell of the cattle, strangely appealing and attractive to those who make it their job to tend the bovine world. The morning drove along in romantic fashion until it came to persuading the cattle to make the steep climb out of Glen Orrin.

At this point some glamour evaporated as we resorted to beating and cursing the stubborn brutes as they milled at the foot of the climb. Eventually a small cut had to be driven ahead to lead the remainder. Sandy stayed below holding the ponies and shouting advice. Once over the high Bealach the herd stepped onto Monar ground. Knowing they were heading for clean unmolested summering they toddled off in good form. The calf born that morning kept up all the way, running tight to its caring mother's side, a credit to the hardiness of a native Scottish breed of cattle. Both a Shorthorn and Angus bull also trundled in the herd. They seemed always in the rear rank of the column eyeing and occasionally grumbling at one another.

Down onto Loch Monarside and over the river at the halfway house. The cattle, thirsty and empty sided, made it clear that they wanted to stop and graze, the whole journey being the best part of thirty miles, with by far the bigger half undertaken that day. We left the tired herd on the slopes of Maoile Choill-Mhias and made for home knowing we could complete the job next morning.

Reaching Strathmore that evening we also felt ready for a rest. The ponies' kit and the Fairburn boys were loaded into a boat and I rowed them over to Pait. Iain and Kenny took a pony each and swam them over the corran. We watched from the rowing boat as the MacKay boys, standing on the backs of the swimming ponies, balanced with the reins. How Sandy wished them to overbalance just by way of retribution for the fire incident. Out of luck, two wet ponies and two dry boys emerged on the sandbank and set off to race their mounts up the road to Pait and supper. Sandy and Willie clumped up the last half mile they vowed to walk that day.

Our drove had yet to reach its destination, the broad grassy flats west of Pait and surrounding the Gead Lochs. Duly next morning the drovers boated down to the halfway house, there to pick up the herd again. Large piles of dung and flattened grass beds told us the herd had not moved much after we left them the previous evening. Counting as best we could, there were 120 cows many with calves at foot, leaving us reasonably certain of a correct muster, and with little difficulty we drove on as far as the corran crossing on the Strathmore side. Here, on the steep bank down to the swim over crossing, fun really commenced. The herd obstinately decided that to the Strathmore glen they must head. Our firm intention was to swim them across to the Pait shore. A battle of bovine perverseness versus sheer human determination commenced.

We summoned reserves in the form of visitors who had unsuspectingly arrived the previous day. Betty's brother Tom and Andy Bole, a friend of the MacKays, armed themselves with hefty sticks. Hector and Alison, keen to join in the melee, we directed to stand at a safe distance. Time and again, howling and beating, we forced the cattle to the water's edge. Equally determined the brutes would wheel round and we were afraid of the calves being trampled. To heighten the fracas the bulls, electrified by the rodeo, decided to fall out in earnest. Head to head, bellowing and pushing, side stepping and rushing, down into the loch they went. We tried to stop them but nothing halted or broke their anger. Soon their noses gushed blood, the water foamed red. All, including the cows turned spectator to watch this awesome display of power and venom. They fought many minutes, heaving, bellowing and bashing heads together with resounding thuds as they wheeled about amidst rock and stone. Belly deep in water splashing and snorting, little wonder the Shorthorn slipped. In a flash the Angus' head went under the rolling belly and using all the rippling power in a hugh neck he literally threw his opponent out into the loch. The poor Shorthorn landed with a wallowing splash. Fear lent speed and before the victor could take further advantage he shot to his feet and fled round the shore.

After this highlight our battle still remained. We went back to the herd which now stood glumly looking at the unpleasant prospect of a swim. "How about the boat?" Iain suggested over the commotion. Something must be tried - I ran round to the pier and took out the *Spray* towing a rowing boat round to the corran channel. Drovers and volunteers alike were now soaked. Betty waded out, I pulled her on board the *Spray* and gave her the job of steadying the boat in position. Those on shore held the herd against the water edge. Taking a long rope out of the *Spray* but leaving one end secured to the stern ring, Kenny fashioned a loop and I rowed him in the little boat towards the cattle. He stood balancing on the stern seat as I held the rowing boat to the beach and threw the loop with a dexterity to be admired, or perhaps luck to be envied, over a cow's head. The panicking animal started back, fighting the tightening rope. "Full speed ahead" Iain yelled to the *Spray*. Betty raced the engine, pulling the choking beast into the water. Watching the straining rope, I put Kenny alongside her and over the gunwale of the rowing boat he reached out to keep her head up. "Beat them on now boys" Iain yelled to the team on shore. Shouts and whacking renewed, the lassoed cow began to swim. It only took one to show the way and the herd began to pour into the water and swim. "Let her go, let her go" shouted Kenny "she's going to drown". Betty hurriedly let go the rope from the *Spray's* stern, Kenny with skill, so he maintained later, got the rope off the distressed beast's head, she staggered ashore on the sand-bank coughing, very much a victim of the system.

To see a large herd of cattle all swimming was a fine spectacle. The cows lunged into the water, heads up, eyes wild and rolling, before they struck out boldly, their fat bellies keeping them afloat like bobbing barrels. Most amusingly, and for what reason I don't know many swam with their tails held straight up in the air. The calves were equally comical, diving in, their heads up, terrified, bawling for their mothers, before swimming frantically as they took a first lesson in an element they only knew as a drink. We wondered about the two day old calf, but again left him to take his chance, he couldn't be caught anyway, and the tough little fellow appeared on the far side as good as the rest. Betty and myself positioned the *Spray* on the east side of the swimming mob to guide them onto the sandbank. They clambered out dripping, each standing a moment to shake itself dry before, fresh and frisky they cantered along the sand to the flats about Pait boathouse Their three day trek was virtually over as the herd picked its way up the road to Pait, out through the yard leading past the Lodge at the green gate, away for the summer to thousands of acres of clean empty country.

Not all summer visitors enjoyed coming to see us at Strathmore. Some thought it too remote and primitive, others didn't enjoy the boat trip if it occasioned a rough one and the worry of their return sail invariably kept them for troubling us again. Only on one occasion did I leave Strathmore for three full days in all those years but possibly three or four times a year I might be away at a sale overnight. Having been away I was never happier than when sailing back up the loch to home.

Betty and the children once went away ostensibly for a fortnight, south to the cities. After little over a week, word came they must return. Betty found the noise and pace too upsetting. The unimpressed holidaymakers returned to Monar on a day that the *Spray* happened to be out of commission. I sailed down to collect them in one of the small rowing boats using an outboard motor. By chance, arriving on the same day, were the MacKay boys and Teenie, their Mother, who had been down country the previous night. Five adults, two children and a pile of luggage and supplies gave a fairish load to a 12 foot boat. So often the case, clearing the east end narrows, we found the wind fresh and the loch quite rough. Teenie, never a good sailor sat in the bow of the boat and pulled a sheet over her head so as not to see the waves. Rounding the dangerous promontory on the south side, long lines of breakers ranged before us, there was nothing for it but to proceed slowly, keeping our head into the wind. Hugging the shore the weather gave us a long wet journey home. The wind caught every bow–full of spray and flung it into my face as I sat at the outboard. The adult passengers sat with coats over their heads whilst Hector and Alison clung on tight, but not obviously afraid. Many would doubtless consider the city a haven of sanity compared to such a home coming.

Only very rarely did people arrive without warning, the area being too difficult to reach for the average hiker. The intrepid occasionally surprised us however. Out on the Pait hill of Cruachan one August afternoon, I came to a spot in sight of Strathmore and sat down to spy the homestead — chiefly I should say, because during the morning I had scythed some grass on the croft for making into hay, and I expected Betty to be working away gathering the green material over to a high wire and post fence which ran down the centre of the field. I digress a little to explain. The purpose of this operation was then to hang on the wires armfulls of grass at a time and leave them to cure into hay. This fresh material, if not put on the fence too thickly would, after a fortnight or three weeks, dry out into the best and sweetest of hay. Spying the croft that gloriously sunny afternoon, Betty was not in sight. Not exactly aggrieved but somewhat surprised, I swung the glass carefully over the distant croft lands. Considerably more to my surprise the glass fell on two strange ponies which appeared shut in the sheep fanks. I set off home a little puzzled, down through Pait. No, the old folks had seen nothing. The boys it happened, were away at the Portree Highland games and I had been out checking the hill cows — a pleasant and undemanding occupation for such a warm day.

I now rowed smartly across to Strathmore frowning a little.

Striding through the trees on the path up to the house, suddenly I paused and gulped. Two large bearded men dressed like cowboys emerged from the porch. I could barely have been more surprised had the outlandish pair been followed by a troupe of performing elephants. They walked down towards me. Unquestionably they might have come off a wild west film set, down to holster chaps and spurs. It crossed my mind this Metro–Goldwyn–Meyer couple might think me equally odd in my plus four tweeds, with stick, dogs and glass, but barely so incongruous I reasoned as they jingled to a halt before me. The taller, more heavily bearded cowboy, lifted his stetson, held out a hand and spoke. "Howdy friend" came the slowest southern drawl. We shook hands. I trusted my good manners covered my amusement — not lest for safety as I noted the unmistakable handle of a colt revolver protruding from a holster. "Say" he continued "ah hope you don't mind pardner, we shut our ponies in your corral". "Not at all boys" I replied, inviting them back to the house for a meal. Apart from the eccentric appearance they seemed sane, charming and friendly.

They made easy conversation and it transpired this hardy pair were touring Scotland on horseback and wished me to put them on the road to Kintail. I explained about crossing the loch, they looked a little doubtful and on going back to the 'corral' I realised why. Balancing on the fence rails were two most elaborate western saddles, plus all their kit, fastened on by various straps. "I sure guess we'd better saddle up and mosey along if you're a saying so". Yes, the evening and the midges were coming on, I agreed. Once saddled up and mounted they certainly looked the part, as down to the corran I led the gun-toting pair. Betty and the children trailed along to watch the fun. Down to the water's edge the bizarre couple rode. Would their horses take to a swim - spurs, curses, lariat ends saw the horses rearing and bucking, everything but into the loch. "Hang on boys, I'll get a pony to lead you across". I ran back to the pier, rowed over to Pait, grabbed poor Dandy who miscalculating the event had also arrived to watch. I threw on reins and a bit which I found in the boathouse and came cantering down, appropriately bare back, for we always rode Indian fashion.

Out along the sandbank, I stood on Dandy's back and swam him across to where they waited and watched. "Say that's a great idea" remarked the bolder fellow, "I sure like the way you stay dry". With Dandy leading back into the water, their ponies willingly followed. Turning, when half way across, I saw the chief cowboy clamber up to stand on his saddle, presumably to keep his six guns dry. Well out and swimming bravely his heavily laden horse unpredictably thought it might drown and gave a lunge. The film set cowboy uttered a southern wail, swaying wildly for a moment, before another lunge shot him gracefully over his horse's back. The resounding splash cut short his howl. Freed of its balancing act, the horse quickly swam ashore, and set off at a gallop up the sandbank. Meantime the cowboy minus stetson floundered onto the sand, water pouring from every inch of his flamboyant attire. "That goldarn orny critter of a nag" he fumed in predictable phrases before noticing the stetson as it bobbed majestically down the loch. "Get that hat" he spluttered. I drove Dandy back into the loch and not without difficulty retrieved the head gear.

His horse vanished, the cowboy now stood bedraggled on the sand bank woefully emptying his high heeled boots. I commiseratingly returned his stetson and set off after the horse. I found the unrepentant nag standing in a corner of MacKays hay field with the fancy star-studded saddle slung under its belly. After some re-organisation, the heroes, with a magnanimous salute set off up the road to Pait. Just like a film-set the sun was going down. "Did the cowboys call in the passing yesterday" I enquired of Mrs. MacKay the next day. "Indeed no, we didn't see a soul" she replied. Evidently the southern gentlemen considered further contact with the natives unwise. For all I knew they might have muffled their horses hooves.

Not all visitors happened to be quite so bizarre or divertingly entertaining. A feature of the remote Highlands over many generations were the tramps. They appeared to travel on circuits arriving from time to time on crofts and houses where they knew tea and a piece would be given. Such a wandering man in the 1920's was the tramp George Munro. A well known figure over a wide area of the Highlands for many years, he came through Pait on his way to the east about once a year. On one particular occasion he arrived at Pait during the stalking season having walked east the ten miles from Iron Lodge in Kintail. A large strong man capable of working when all else failed, he asked for a job. The ghillies, as it happened, were busy carting in peats to stock up the Lodge with fuel for the forthcoming shooting season and Munro was instructed by them to have a meal ready that evening. In due course the hungry workers repaired to the bothy tired from their day at the peats. To their delight a large pan of soup and another of meat steamed invitingly on the open fire. Munro looked pleased with his efforts and made a show of serving the repast. However some of the ghillies, not generally given to fastidious affectations, were put off by the meal's extra strong, not to say its decidedly peculiar smell. Hunger quelled qualms and all partook of a plate of soup remarking ungratefully upon its queer taste. In unison the diners refused point blank to touch the evil smelling meat. Munro seemed hurt and offended at their churlish disdain for his efforts. He took hearty helpings himself and to emphasize the savoury goodness of the main dish partook of a second helping of meat. Later that evening the ghillies felt dreadfully ill. Vomiting and diarrhoea swept through the bothy. A night of misery ensued, the most grievously afflicted fearing death at hand. George in perfect health sympathised with the stricken band. Under threatening enquiry the itinerant cook admitted that on his journey over to Pait he had noticed a dead hind floating in the Gead Lochs and thinking to give the boys a treat when asked to cook a meal, had hurried back to retrieve the delicacy. Several of the ghillies on learning the truth were sick again, and remained indisposed for days.

Hay for winter sheep feed came up each summer. The consignment allotted to Strathmore I used to store in a shed about 50 yards up from the lochside in the wood at Reidh Cruaidh. Carrying up the bales of hay from boat to shed was a chore we did not relish and when word came to expect a load at Monar in a day or two, a plan to ease the burden took shape in our minds. It so happened that a troop of adventurous Manchester Boy Scouts were trekking through the glen and asked to camp a few days in the wood surrounding Strathmore Lodge. We made them welcome, greatly enjoying their camp fires, songs and company. They in turn were full of interest in our life style. "Would they like a trip on the boats the following day?" we pleasantly enquired. The Scout Master enthused "What a splendid idea, but was there enough room?" "Plenty, plenty" we assured him. The pleasure cruise was arranged for the next afternoon. For years the craft used in transporting the hay happened to be an ancient and dilapidated flat bottomed salmon coble of extremely doubtful sea worthiness. She had been lying out of the water since the previous year's operation and the boys and myself launched her at the 'new' boathouse that evening, hoping the wood might swell a little overnight.

Next afternoon the helpful weather turned hot and sunny, beautifully calm, a superb day for a sail. After considerable bailing, we towed the hulk round to Strathmore pier and loaded the eager Scouts. Sailing down to Monar in carnival spirit the day shone to match. Shirts came off. Scouts, landscape and scheming shepherds basked in sun and balmy air admiring the reflections cast in the waters of a perfect day.

Striking a note of caution we stationed young Kenny in the coble, surreptitiously equipped with a pail but the old tub riding high out of the water, towed easily and obediently. Long curving ripples fanned out astern, our gliding voyage broke the vain scenery's silver mirror. Thirty Mancunian Ulysses bowed in enchanted obeisance before the sheer magnificence of the harmonious grandeur of nature, personified through our Highland scene.

Waiting on Monar pier as we drew alongside were 300 heavy bales of hay. The day clouded a little, and as we set about loading the coble, the Scouts, being trained in acts of social welfare, gave us a hand. Kenny took responsibility for stowing the hay on board the coble and ensured a large space was left at the stern. The load grew to seven rounds of hay carefully stacked by sweating scouts. When loading was completed the exhausted coolies perched on top and stretched out. It made an unwieldy, if not dangerous load. We set off immediately, Kenny stationed apprehensively in the cargo space gripping a pail. For some reason that day the *Spray* was out of service and our tow boat had to be the old An Gead. We tried to pick up a little speed, the hulk well below her plimsoll line began to yaw in a most unhelpful fashion. We had secured her with heavy ropes passed to the two stern posts of An Gead. First one and then the other post would get a nasty jerking tug. Privately knowing the state of her timbers, we began to fear for the stern of the old launch. I looked at Iain, he at me. We both turned to view the perambulating barge with mounting unease. The Scoutmaster sitting with us, as befitted a V.I.P. began to look concerned.

We were well out into the middle of the loch when the first pail full of bailing water shot out from the rear of the quietly sinking coble. The Scoutmaster turned at the splash, paling through a newly acquired tan as the implication of Kenny's now frenzied activity caught his imagination. The speed of bailing increased. Iain changed course and moved across to the north shore. "Is everything alright?", the ashen Scoutmaster gulped, his Adam's apple yow-yowed alarmingly. "Och aye, she's just taking a little water being down at the bow you know. Kenny will keep her right". Kenny by now we knew would be sweating like a Bombay stoker. Worse, violent tugs on the stern of the tow boat opened a seam, and water appeared over our own floorboards. I began to pump at once.

Two sinking boats and thirty scouts spread over six tons of hay suddenly became an uncomfortable liability. To cheer things up and break the tension I commented lightly that we trusted the Scouts had all passed their swimming badge. The Scoutmaster not heartened by my humour stood up, looking wildly about, clearly judging shore distance against the rate of bailing and pumping. Wringing his hands he merely glared in reply.

"Don't worry boy, it's just round the next corner". I unwisely put a hand on his trembling shoulder. "You'll answer for this" he hissed shaking me off. Rounding the headland we chugged sluggishly like victims of a bombardment over the last stretch to Reidh Cruaidh. Now thoroughly water logged, we grounded well out from the landing point and hay shed. Indicating there could be nothing else for it, and they were going to get wet anyway, we suggested to the Scouts they form a human chain from this side of the boat through the water and up to the shed. Out of sheer relief they did as we suggested. The hay went into the shed in no time. Wet and weary, the Scouts declined the remainder of the sail, preferring a two mile walk back to Strathmore. No invitation for us at the camp fire that evening. We appreciated of course they would be tired. Next day they trekked off, pulling barrows and carts out over Bealach Bhearnais and down to Achnashellach. Hardy boys.

Few people penetrated our wilderness. Those who did so generally knew something of mountaineering craft and did not underestimate the distances and possible dangers. Lying reading on the couch in the early hours of a morning two or three days after New Year, I felt snug and content. The paraffin tilley hissed out a good light, the Nu-Rex stove gave a cosy heat. Outside the heavy snow of the previous day had given way to hard frost. A tap at the window surprised me. Rising from the couch I stepped over and parted the curtains. Half a dozen pinched faces looked in. I opened the door to find eight young chaps truly glad to find even our primitive civilisation. Taking them inside at once, I called down Betty who had not long gone to bed. We stoked up the fire which they eagerly huddled over as Betty made porridge and then stew for them on our Calor Gas stove.

Their experience unfolded. That morning they set out to climb Sgurr a Chaorachain from Achnashellach. After a slippery ascent, blizzard conditions struck on the top and they had become disorientated, in fact completely lost. At least they had the sense to bunker down to allow the weather time to clear before, with difficulty and danger they came down from Bealach Toll a Chaorachain cutting their way in ice steps. During the day I had been watching the snow blowing like white smoke on the Strathmore ridges and guessed what they must have suffered. Several of them slept after the meal, one lighter clad chap I guessed close to frost bite in his hands and feet. He suffered much pain as circulation returned. They were a party of students and confidence restored, became anxious to get back to base before the alarm was raised. At about half past four I led the weary party up behind Strathmore and through Strath Mhuillich to the Bealach which set them easily down toward Achnashellach. Lucky chaps, who by now realised what a risk the day had involved, bid a thoughtful farewell.

Not so fortunate at the turn of the century was a tramp who passed through Monar on a March day heading for Strathmore. In Highland tradition the then Monar keeper, one MacDonald to name, saw the vagrant fed at the house before the ill–clad man set off west in the early afternoon. The day blew a sudden Spring storm of great severity and nothing more was heard or thought of the tramp. Late that April a shepherd walking west to Strathmore noticed his dogs behaving in a peculiar way about the remains of a snow drift. He investigated and there found reasonably preserved human remains. It turned out to be the body of the tramp whom nobody had missed or even enquired after. In due course the ghillies had the unenviable task of taking the now decomposing body down to Monar on a tarpaulin. They gladly deposited it at the boathouse whilst the authorities debated who he was, where he should be buried and above all who should foot the bill for a coffin? Meantime the disgusting but true story continued: rats attacked the remains and the ghillies were directed to sling the sheet from the boathouse rafters to prevent further unpleasantness. No record of his identity could be found, perhaps only the man himself might have known. After several weeks the noxious bundle was taken away by the police.

Visitors, none quite so unlucky, came and went but our summer work fell into a definite pattern. During June our main work with the sheep stocks became the first of the extensive gatherings for lamb marking. Each hirsel, gathered from the hill in turn as the weather permitted, was driven into the fanks for handling. The basic plan of our sheep pens or, as we called them, sheep fanks, led from one main collecting enclosure to either a drafting race or to various smaller pens. Out of the large main pen the flock would be forced through the drafting race, a narrow blind sided passage which we called a 'shedder'. The shepherd stood at the end of this arrangement 'shedding' the oncoming sheep in three directions and therefore three groups of sheep could be separated by the operation of two gates.

Hung opposite to one another they swung so that the passage could be turned to either side or thirdly, when both were closed, the route lay straight ahead.

We used the common practice for dividing the flock prior to whatever husbandry operation followed. Much 'woofing' and shouting helped to bolt the sheep down the shedder. The operator had to be smart with eye and gates to make a clean shed. The dogs relished a day at the fanks, it gave opportunity for sly nips at the sheep which stern shepherds would never tolerate in any other circumstances. We did not allow young dogs to work at the fanks, as too many bad habits were easily learned. Sometimes also the sheep in a stampede, might run over a young dog starting to work and this could unnerve the animal for life. Shep, anything but unnerved, was an ace worker at the fanks. Combining barks, bites and boldness he forced the flock without mercy, pretending to look guilty and ashamed if I cursed him for having a tuft of wool hanging at his mouth.

Apart from the old stone fank halfway up Strathmore glen which we only used as an overnight holding pen, our fanks were of wooden construction and built to a celebrated College design. The arrangements had some shortcomings as Sandy in fury remarked on one occasion when driving the sheep into the stupid circular dipper pen. "The so and so who had this brain wave should be made to work it like a tread mill for the rest of his life". The old style fanks were all stone walled of course, and it was the fashion of the dogs to run about the turf covered wall tops barking down at the sheep. In short, our's was 'modern' but had frustration built into its plan.

It took a good hour and much scooting about on a June day to see the shedding completed, and a bleating penful of turmoil result. Our first task was marking these scrambling lambs. Each squirming lamb had to be tailed, stock marked and the males castrated. The latter were caught and held and the operations deftly carried out using a clean sharp knife. The tails swiftly cut off about the level of the hock didn't cause much bleeding and gave a greater degree of hygiene in years to come for the ewe lambs, whilst better fattening potential, many claimed, for the wedder or castrated male. Essential stock marking of the ear for identification on our open unfenced hills came next. The Strathmore mark for over a hundred and twenty years was a forked cut in the right ear. This I carried out by folding the ear and cutting off the corner from the back.

Castration required a bigger operation. The lamb had his legs held by one operator so that the scrotum stood out. I then cut off the bottom of the bag in one quick stroke which exposed the white tips of the testicles. The shepherd, so long as he did not have dentures, caught each testicle in turn and taking a firm grip between his teeth, held the scrotum with his fingers and pulled them out. The testicle came away quite freely taking a long length of cord with it. Waiting dogs snapped the testicles up as the shepherd spat them out. A dash of healing oil quickly applied to the open wound saw little bleeding if the technique were expertly carried out. Off the lamb went, now a wedder. Though the exercise looked uncouth and primitive we always considered this method of cutting preferable to the use of rubber rings. This system, developed in New Zealand, came into fashion about that time and caused the lambs to writhe about in pain for many hours after the application of a band to the top of the scrotum.

After these first most necessary tasks of the sheep husbandry calendar, the ewes with lambs returned to the hill. We kept in the hoggs and yeld ewes for the first of the season's clipping. In a ninety inch rainfall area wet days often plagued us. Every opportunity had to be grasped to get the maximum achieved on a good day. Clipping I always found a fine, satisfying job when the day was good. Speed, rhythm, skill and a good honest sweat made heavy work seem a pleasure.

Most of the glen clippings worked on a communal basis. For the main Strathmore clipping, towards the end of July when all the ewes with lambs went under the shears, we would have a notable line up. Sandy, glad of a break from Fairburn, would bring his grand dog Moss. The children thought it the best day ever, running about excitedly asking to help. We employed them to put the paint mark on each sheep and oil on any cuts which the shepherd might make. In half an hour clothes, faces, legs and hair became covered in red marking fluid. Big Bob walked over from Benula in Glen Cannich and combined impressive clipping ability with a fund of stories. Naturally the MacKay boys gave a hand, whilst old Kenny, then nearing seventy, rolled the fleeces and managed his pipe at one and the same time.

The fank became a hive of industry. We all clipped using hand shears, in the modern style of working the sheep on the ground, as opposed to the old way of placing the ewe on a specially shaped clipping stool and tying her legs. The former method, though quicker, did not produce such a good result. Turn about was taken at catching the ewes out of a small pen and presenting the sheep to the clippers who, after elaborately sharpening their shears, stood over a spread wool bag ready to go. We soon had old Kenny rolling fleeces for all he was worth. Betty between making tea and meals came out to give old Kenny a hand. If the day turned hot, shirts peeled off, sweat poured down browning backs. Amidst the constant metallic snip snip of shears flowed news and banter giving us a happy neighbourly day. Sometimes we clipped on late, though invariably the midges forced a stop, but three or four hundred sheep might have passed under the shears. The day ended with a good well deserved dram of whisky.

Throughout July, gatherings or clipping occupied our days, the work being hard and strenuous. Sundays in the summer afforded a welcome break and alternative weeks, Pait or Strathmore, we made our tea with the MacKays. Stories told of generations ago never wearied us. Mrs MacKay possessed an unending fund of local anecdotes and also much old Highland folk lore, which she related in an amusing way. She had a typically subtle Highland sense of humour which left the more dubious passages to the imagination, gaining much wit simply by inference. Naturally anything of local moment provides conversation, and afternoons and evenings in their company passed quickly and pleasantly.

It could be into August before the meadow hay on the croft would be ready for cutting. Strenuous efforts in May and June to keep out marauding stags meant constant vigil. I used to tie villainous Bob to the fence but the all night barking got us down. A light worked for a time, then had to be changed to a waving flag tied on a string. In short any ruse sufficed, provided it caused the hungry brutes to be suspicious and remain away from the precious crop. Old Kenny told of the enterprise of one stag many years previously. His hayparks lay below Pait on the flat between the Riabhachan burn and the Uisge Garbh. He kept them secure with well maintained deer fencing. Much to his annoyance late in June one year, heavy tracks and droppings in the park indicated that an intruding stag was helping himself to the essential hay crop. The fence appeared unjumpable and unbreachable. Kenny was both annoyed and mystified. The damage increased and must be stopped. He decided to stay overnight in the old barn above the pier which commanded a view over the fields. Daylight filtered in on a raw morning as a troup of stags came grazing down to the fence and commenced to feed along its boundary. Unconcerned, and thinking themselves secure at that early hour, they crossed a deep drain which ran through the croft. In fact it had once been the old bed of the Uisge Garbh burn. As soon as the band were all across, one heavy stag turned back and going into the water of the drain swam under the fence, thus gaining the entry which had so puzzled Kenny. Needless to say, as the artful stag grazed up the field towards the barn, he paid the price of his cleverness.

I cut about an acre and a half of meadow hay with the scythe. Using the hand barrow Betty, with the children trying to help, carted the freshly cut material and built it on to a specially strong high fence made for the purpose of drying the grass. Without too much bleaching this laborious but foolproof method made a first class palatable product. It required some three weeks wilting on the fence, then given some windy days it came ready for the barn. Again the hand barrow became the means of transport. I piled on big loads, threw a string over for keeping it on, and always succumbed to the children's pleading for a ride in the hay up to the barn. I suppose only about two tons was made but together with the heather, potato peelings and a little bruised oats, the cattle came through the winter remarkably well. They gave us a drop of milk right on until six weeks before calving.

The estate hill cattle, only handled once during the summer in the Pait fank for castrating the bull calves, otherwise enjoyed themselves with unlimited free range. Occasionally they wandered away west to the Attadale ground at Ben Dronaig, there to mix and no doubt fight with a neighbouring beef herd. One evening as the sound of such bedlam drifted down to us, we trekked out with dogs to give ours a good hounding back. Later in the summer when it suited their needs they wandered round to Strathmore. Once the middle of September came in, I had to watch each day to keep them on the west side of the house otherwise off they would set away down to Fairburn. Whatever their method of assessing a problem their activities were by no means stupid and I often wondered which of them did the thinking or whether it might be a corporate decision. The cattle looked after themselves remarkably well, and I recall only isolated accidents.

Going out to move the herd to Strathmore one September we found a fine blue grey in-calf heifer lying with a broken hind leg. We left her behind at the west end of the Gead lochs as we pushed the herd along the drove road and round the east shoulder of Meall Mhor into Strathmore glen. She stayed alone. By November though we had left her quiet, apart from taking up a little feed every other day, the leg remained unhealed. Instructions from the estate arrived to take her home to Pait. Iain and I set off to drive her down but the beast could make little headway in the rough ground. Eventually she staggered into the water at the west end of the Gead Lochs and stood belly deep, looking balefully at us. High on the bank at the other end of the lochs lay a rotten old rowing boat. We left the beast and together went down for the boat. Iain rowed, I bailed smartly and we eventually reached close to the heifer still in the water and now watching suspiciously.

Thoughtfully that morning we had taken a long rope with us. In a clever throw Iain lassoed the poor animal. It floundered out into deep water. Iain began to row hard down the loch. As well as bailing I pulled the heifer, now floating at the stern of the boat. Thus we set off towing her down the Gead Lochs. With the sun long down behind Meall Mhor, frost began to set in as we beached the wooden sieve of a boat at the east end of the lochs. The exhausted animal dragged itself out over the stones and there had to be left. Sure to find it dead after the cold night we went up the next morning to be astonished in discovering it halfway down the path to Pait. After two days the unfortunate beast made its halting way into the yard and old Kenny tied her in the byre for the winter. The heifer's leg eventually healed by the spring but stuck out at a hideous angle. On hearing this news, further instructions came to shoot the beast and send down the meat. Iain shot her but the carcase made only dogmeat and the estate used it as such. The end of an unfortunate animal. We had wanted to shoot her the day she was found injured.

The last year we were in Strathmore it was difficult for me to keep the hill cattle from making away to Fairburn and do other work at the same time. By mid-October the herd became determined to head east and one particularly cold day of sleet showers it seemed their minds were firmly made up. I decided myself and the family should sail down to Monar at post time in the *Spray*. Our second launch, the *An Gead*, lay at Monar pier and I wished to tow her home. It had been an autumn giving us six weeks of dry weather with east winds, resulting in the loch falling very low. Of necessity I had the *Spray* lying at anchor east of the corran, the depth being too shallow to allow her to come over the sand bar. We rowed out to the *Spray* as the wind freshened, cold and yet again from the east. Down to Monar we sailed, gaining a little shelter behind the cabin from a nasty snatching wind. Betty took over the *An Gead* as I towed her back up the loch. The wind blew to a gale with heavy bitterly cold sleet showers. The *An Gead* had no cabin or shelter and Betty began to get cold. The children huddled down under a tarpaulin in the *Spray's* open fronted cabin. We reached the sandbank with big rollers following us from the east. I dropped the *Spray's* anchor, shut off her engine and as she swung round the chain Betty called to me "the cattle are making past the back of the house".

Looking up I could see them high against the whitening hill-side marching determinedly eastward into the sleet. Time to shape myself I thought. The *Spray's* anchor was not sufficient to hold the two boats in what now blew a full gale. I visualised two boats washed up on the sandbank and the cattle away; it spelt real disaster. Only with difficulty did I heave the *An Gead* close enough to the *Spray* to let Betty lean across. The wretched anchor pulled with each wave. I handed over the children to Betty whilst trying to keep the boats from damaging each other as they plunged side by side. Jumping into the *An Gead* myself, I cast off at once hoping now that the *Spray* alone on the anchor would be held. Free in heavy waves *An Gead* began to drift on the sandbank. I fumbled with the engine, it wouldn't start. With an awful crunch we were aground. The pounding and rolling could easily capsize the boat. The children's cold faces now looked frightened. Swift action. I went straight over the side without time to consider the freezing water. Waist deep I heaved, Betty pushed with an oar and bit by bit we moved the launch down to the channel. Shouting to Betty to push the bow round, I put my back to the stern and used each wave to heave her over the shallows into the calmer head waters. Directing Betty to get her round to the boathouse as best she could I scrambled ashore, emptied my wellingtons and set off at a run to head off the cattle.

It was now almost dark. I felt my limbs coming back to life in the half mile up-hill run to intercept the fleeing cattle. Stumbling and cursing I reached the leaders as they were heading doggedly along the top of the hill park fence. Their coats were blankets of snow, their eyes like black goggles, but without mercy I drove them to the west side of the Strathmore croft fencing. Trusting it might hold them for the night I hurried back past the house. At least with the tilleys going I knew all was well and didn't stop. Back down to the sandbank. The *Spray* rode too close in for safety, the trough of each wave was letting her keel thud on the sand. Nothing for it, being totally wet anyway, I waded out to her and pulled myself aboard. Only then did I realise how tricky it was to get into a boat from the water and after falling back twice I clambered with difficulty up the rudder. The engine started without a fault. I drove her slowly ahead, whilst moving quickly to the bow to lift the anchor. Taking a fresh position twenty yards of strong chain rattled out. After waiting a few minutes to check that she held, I went over the side. This time I was really cold, and oddly the water seemed warm. I swam a few yards till my feet touched the bottom and waded thankfully ashore. Hot soup and a large meal of venison waited for me when I got in. Next morning the cattle had

vanished, having trekked out during the night in spite of my efforts.
Wise beasts, they knew better than I the halcyon summer days were past.

THE PAIT BLEND

Driven rain pelted the windows with staccato volleys and rattles, driven in a fierce blast that paused only to sob as the elements drew breath, to hurl themselves renewed with Herculean force against our puny human foothold in their rightful domain. The fire scudded and shuddered, the gale roared over the chimney head, angry that a stone gable attempted to block its freedom. We sat in close to the well stoked fire, the Tilley lamp hissed from the shelf; both added friendly reassurance to the tiny kitchen which protected us from the wrath of the outside world. My feet, heels singeing, rested on top of the stove, matched many evenings by Iain's pair warming at the other edge. We sat snug and talked often about old days and ways in yesterday's Highlands. Betty's needles grew another sock, or jersey, for the children, long since fast asleep below the clattering slates.

The hardy resilience of our predecessers who stalked or shepherded these far domains was exampled by two singular men. I quote from William Collie's book of Memoirs. Collie, a Keeper at Coulin, on his first visit to Monar, and having walked into the ground at the west end, met both these men. 'Half way down the Strathmore, we met two giant shepherds, Hector MacLennan (Eachin Mhor i.e. big Hector) and John McGileas - typical sons of the north. Eachin, standing well over six feet and proportionately framed appeared an imposing figure in his rough Highland garb. His kilt consisted of only a yard and a half of cloth, with three broad plaits across the buttocks. Ordinarily, nine yards of cloth or upwards is pleated into a full kilt. A

sporran made from deer's skin, a thick tweed jacket and vest, a huge
Tam o' Shanter bonnet with a massive red tassel, strong, heavy
home-made shoes, and stockings of wool spun and knitted by his
wife, completed his outfit and he carried a large shepherd's crook
which measured almost his own height.He walked with a long elastic
step, erect carriage and free bearing - quite a noble specimen of a
Highlander. His fine large ruddy features were however slightly
marred by an accident to an eye which was rendered sightless.

Each lived in a house built by himself, typical of the times and
district. 'There was no external display, but inside every comfort, as
they best understood in such circumstances and surroundings, was
provided for. The fireplace consisted of a raised platform a foot high,
built in the middle of what might be called the living or sitting room;
no grate and no chimney, the smoke finding exit through a small
opening in the roof or through the door or windows as it best could.
Houses of this description were then common enough in the West
Highlands of Scotland and are still to be found in the north. In
disposition, as in appearance, Eachan was in many respects a
wonderfully interesting character. He could not converse in English
with any degree of fluency but his Gaelic, spoken with a soft and rather
refined tone, was perfect'.

Collie came to Monar from Coulin in 1861 as head stalker and
sheep manager when the Estate was purchased by an English laird, one
Mr Holmes. The Stalker's house was situated on a prominent knoll
high above Monar lodge, whilst to the rear of this house ran the same
noisy burn which hurried past the Lodge. The surrounding trees
sheltered the croft land and steading, perquisites enjoyed by the keeper
as part of his emoluments. He kept a small number of cows and the
few acres of arable were cropped in rotation - oats, potatoes and
turnips. As sheep manager, Collie employed Eachan Mhor and his
brother John to shepherd Strathmore, no small task for them with a
stock of 3,000 head. The handling facilities in those days were at
Luib an Inbhir, halfway between Strathmore and Monar at a point
where the fourth largest of the six main rivers which flowed into the
loch spread wide green alluvial flats.

Here the great clippings took place. To facilitate the work, a
galvanised iron roofed shed, capable of holding 400 sheep at a time
was erected by Collie for as we knew only too well, rain very often
interfered with the shearing. Locals far and wide would foregather.
Each household, large or small, had a young girl as a domestic servant.
The lassie serving the Luib an Inbhir family would have been

despatched, prior to the clipping week, for messages to the shop at Allt nan Sugh three miles below Killilan. She crossed by the sandbank on through Pait and Corrie Each to the shop such as it was, returning the same day with the much prized tea and sugar. A round trip of 35 miles and two wettings at the Corran.

Collie's wife, superintending the commissariat whilst the shearing progressed in full swing, ensured mountains of venison and potatoes. Collie himself looked after things generally, but kept a specially close eye upon, as he recounts 'a hog of whisky, of which I was not by any means sparing – ever having in mind the fact that to get good work out of the men they must be kept in good heart. At the close of the day's work, I invariably regaled them with an extra big bumper'. The company had great fun with Eachan, he could be heard almost from the top of Beinn na Muice shouting and whistling to his dog 'Glen'. It was rather difficult to catch strong sheep on soft muddy ground. The shearers often fell in the act of catching, but Eachan never once lost his footing and easily gripped more sheep than any of the other men. He had a habit of carefully buttoning his coat right up to the throat, evidently thinking that this would afford some protection should he happen to fall. There were few his equal in shearing and smearing sheep.

The method of clipping in those days required the sheep to be lifted onto a long four legged stool. The clipper sat at the narrower end with the sheep stretched out before him. Its legs were tied with a leather thong. A barrel of Archangel tar and butter mixed together sat at the shearer's hand. The long clip marks lay parallel to the length of the sheep's body, and each deft stroke revealed more of the soft white blanket of fleece which spread over the stool as the clipper worked. The fleece fell clear to be collected, rolled and packed by the remaining team. The clipper reached to his tar barrel and under each fold of remaining wool smeared the sticky mixture both as water proofing and to rid the body of parasites. The scene must have been one of hard back-aching work but no doubt helped along by teasing and witticism. No local, it seems, commanded more native wit then Hamish Dhu (James MacRae) who in addition to his lively tongue featured as a specialist in what he and many besides considered the honourable profession of illicit whisky distilling.

Hamish Dhu and his sister Mairi lived with their father Alexander (Alister Mhor) at Pait, a placename said to signify 'The Hump' (of Monar). In fact their little thatch roofed house sat securely situated on what in those days, the 1840's, was an island. In the style of the times the house combined two main rooms, byre and hay shed in one line. It sheltered behind a green hillock of ground picturesquely overlooking the head of the loch between two branches of the Garbh Uisge burn. A certain area of tillable land, rock strewn but mineral rich, lay around the homestead. When not engaged in smuggling, Alister worked this small farm or croft from which, though he derived a fair income, there can be no doubt whisky distilling constituted his main enterprise. Stone tumbled remains of a deer 'watcher's' house called Cosaig, built during the shooting tenancy of a Mr Winans in the 1860's, were to be found on the green banks of a small bay on the south side of Loch Monar. Close by these ruins, some 200 yards up the burn, discreetly nestled a now ruined 'bothan' one of Alister Mhor's early business premises. From this location of Highland enterprise Alister, early in his career, by some inadvertant miscalculation was apprehended by the 'Gaugers'. A large gloating posse proudly marched the prisoner to Dingwall. No stigmatism attached to the crime in local minds but the shame of apprehension dealt a keen blow to prestige. Vowing not to be taken again Alister relocated the site of his venturous operation at the west end of the loch and built a commodious bothy effectively concealed amongst undulating hillocks.

Alister Mhor, noted throughout the district for his great strength, was looked upon by many as the most powerful man of his time and his friends in the trade lost no opportunity of advertising his wonderful feats, hinting also that he turned a terrible savage when roused – propaganda designed to inhibit the enthusiasm of the zealous excisemen.

Alister and his son Hamish spent the winter months hard at work distilling whisky. Strictest secrecy was essential. The bothy carefully dug out of a low hill on the east end of Meall Mhor was their abode of operations. Difficult to approach due to it being surrounded by peat bogs and marsh land, I found their headquarters one day after a long search. Soundly constructed of stone it had ample space for the conducting of a sizeable professional industry. Thus father and son repaired to their premises during the winter, working according to the dictates of their chosen calling. For fear of being tracked to the den by excisemen or other enemies they never ventured away from the bothy when snow covered the ground. Justifiable caution allied to dedication for the quality of product saw weeks frequently pass without their being able to visit home.

The nearest excise depot was stationed amidst an infamous area at Beauly, a rough forty miles from the whereabouts of the copious productivity from MacRae's bothy. On account of the propaganda circulated regarding the uncertainty of Alister Mhor's temperament, the excisemen manifested natural chariness towards venturing into his territory. Indeed Collie records 'he (Alister) was 70 years of age when I first met him and even at that time was capable of showing extraordinary strength'. However at 76 Alister's health began to fail and the work at the still devolved upon Hamish. The old man still usefully occupied his days sitting on a spying stone which he arranged into a comfortable chair perched on a commanding knoll high above the Pait river flats. With his 'prospeck', an old word for telescope, the veteran smuggler would scan the loch down to Monar checking all movements. I found Alister's stone chair on a hillock, reputed also to be the burying ground for infant fatalities in the locality before the days of registration.

With the enfeeblement of Alister, the excisemen became bolder. They were less afraid of Hamish and determined to bring an end to the audacious activities of the MacRae's. Official glory would devolve upon any excisemen who would bring the notorious couple to justice. Not unexpectedly the local people were friendly disposed to the smugglers and regarded the safeguarding of this most necessary indigenous traffic as thoroughly laudable. Indeed when two important and purposeful officers appeared at the door of old John Ross's home at Ardchuilk, half way up Strathfarrar, he quickly surmised their intentions and insisted they stay for dinner. After a heavy repast he charmingly entertained the pompous pair, all the while plying them with the vile duty paid whisky dispensed by Urquhart's Struy Inn. So freely were they regaled with story and drink they remained with their captivatingly plausible host until the early hours of the morning. Meantime a ghillie despatched by Ross arrived at Monar where William Collie, alarmed for the safety of the free enterprise distillery sent two men rowing up to Pait. Hamish, adequately warned, quickly hid all his whisky and distilling utensils.

The excisemen emerged doubly eager to continue with their mission and boastfully resolved upon the capture of Hamish and his father. Striding out as best they could for a further two miles, they suddenly succumbed to the effects of John's hospitality. Taking refuge in a wood beside the road, they fell into a moaning slumber only to awaken after darkness had set in. A day had passed. Their ardour became somewhat dampened. A decision to abandon the hunt was solemnly taken and the woebegone pair sneaked back to Beauly in the dark. Later to regain dignity they regaled their friends with stories of the terrible sufferings experienced in the wilds of Ross-shire. The noble cause of ensuring Highland law and order was upheld Collie observed, as he praised the efficacy of the obnoxious blend 'indeed Urquhart's Highland Dew is just the worst whisky I was ever forced to drink, but look here, everything will have its use'.

Alister lived on to his 97th year never the worse of an attachment to the product of his calling. In the harsh damp climate in which lived the people of the glens, often inhabiting the most humble and primitive accommodation, whisky, and its ability to warm the body, was often a necessity. Hamish netted a considerable sum from his illicit craft and felt proud of his adroitness in evading a tax which he honestly believed should never have been levied. Like most people of the district, he felt convinced that no dishonour could possibly attach to anyone who smuggled whisky.

Hamish Dhu retired eventually to the Old People's Home at Aigas, Kilmorack. His mother however had died many years before in their island home at Pait. A strong matronly woman of ample size, she lived all her married days at Pait, but as a native of Kintail her remains were to be carried back to the parish that saw her birth. Out of due respect and for some perhaps of less charitable inclination, anticipating a good funeral, a sizeable cortege assembled at this outlandish spot to convey the substantial remains of the lady to her last resting at Clach an Duich in distant Kintail. With fitting deference to the solemn occasion, the procession wound its way through Pait, west along the Gead Loch side and into Corrie Each. The cortege bore on gravely and with dignity. Some were employed as coffin bearers, others, more importantly on such an arduous trek, to convey the victuals, contained mostly in large wooden kegs. The day grew warm. Certain ungracious comment, regarding the deceased lady's weight filtered through the ranks of the sweating bearers. At each coffin resting point along the path, stones were added to an existing cairn, or a fresh cairn was piously raised. The warmth grew oppressive. The procession grew thirsty. By and by the frequency of the stops precluded dignified cairn building obsequies. Before the final steep climb out of Corrie Each the Kintail men, waiting high above on the march to join the oncoming mourners, observed below them, formality dispensed to the point where brow mopping bearers seated on the coffin were libaciously quaffing unwise amounts of the otherwise ample supplies. One might consider it befitting that Mrs MacRae, to quote common parlance, was enthusiastically afforded "a good send off".

The local landlords turned a blind eye to MacRae's central

enterprise. Mr Holmes of Monar in the late eighteen hundreds openly bought supplies from Hamish. On the other hand John Stirling, Esq., of Fairburn who purchased the Monar Estate from Holmes adopted a more circumspect view but did not take any measures to dissuade the freebooter. Hamish legally enjoyed squatters' rights. He lived on an island, considered himself a free man and with Highland arrogance "equal to all men and superior to most". Certainly few could equal him in dress or appearance. On fine Sundays MacRae would don the kilt and cocked bonnet for a visit to Monar. An Englishman called Page who spent the summer pike-fishing at Monar would also dress himself with kilt in vainglorious arrogance. Nature had not been kind to Page; he lacked the build to wear Highland dress. This ensured that all who witnessed the pair strutting together on a Sunday morning in the grounds of Monar Lodge relished much quiet amusement.

The Government reward of five pounds for information leading to the locating and capturing of a still came to Hamish's ears. Sometime afterwards, possibly being persuaded by the Stirlings, or more probably as he was tiring of the trade, MacRae dismantled his bothy and buried the still. A journey to Beauly market following a suitable interval and MacRae dropped a quiet word to the excisemen, letting them know that he might be able to help in the discovering of a still if the reward would be forthcoming. It needed only a gentle hint to bring a 'posse' hot foot up Strathfarrar and west to Pait. Gravely innocent, Hamish led them to a peat bank, here making pretence of a puzzled searching, then with an exclamation "the rascals, what did I tell you?"he proudly pulled out his own copper pot and worm and graciously accepted the sum of five pounds for the well worn utensils.

Largely due to the blatant activities of the squatter, MacRae's island sinecure had become a little irksome to the Pait Estate owner. Pait forest at that time ran in with Kintail and the vast area was keepered by Theodore Campbell. About the turn of the century on instructions from the estate, the diversion of the Garbh Uisge by the building of a causeway and revettment bank was undertaken by Campbell. MacRae's island home became joined to the mainland and prevented anyone following the wily Highlander's example. About the same time similar work deflected the Riabhachan burn further east in an effort to deepen the Corran for navigation. Some few years later Tommy Fleming the Monar keeper, this time working at the east end of the loch, partially dammed the mouth of the river Farrar as a result of which loch Monar's level rose by over a foot. The same intention of improved navigation at the head waters was the aim of their laborious and heavy hand work.

Theodore Campbell keepered at Pait around 1898. His father Colin had lived in a tiny 'deerwatcher's house' half way down the south side of Loch Monar at Cosac. This was the march between Pait and the Lovat Estates. Theodore, a large powerful man, apt to be sharp with incompetent ghillies happened at a stalk with a young gentleman to shoot two stags high up in Corrie Riabhachan, a good four miles from the larder down at Pait. Late that evening a crestfallen ghillie arrived at the scene of the gralloch to inform the stalker that the ponies had bolted. Campbell flew into a rage, threw the nervous ghillie a drag rope, "Follow me", he instructed the frightened man. Taking a rope himself Campbell, using his great strength, dragged one of the stags home by 1.30 in the morning. May I assure those who have never dragged a stag over long distance or any distance for that matter, this represented a singular feat of strength. The exhausted ghillie arrived two and a half hours later, claiming boldly to his fellow bothy mates after he recovered that of course the Stalker had pulled the lighter stag

Many Estates from the 1840's onwards were converted into deer forests. Wealthy shooting tenants spent their industrially -produced fortunes in the new vogue of deer stalking. Certainly the most notable and possibly spendthrift tenant at this time was the wealthy American Mr W.L. Winans. His seemingly endless funds were supplied by the profits he made from railroad construction. Coming to the area in the 1880's he rented Glenstrathfarrar from Lord Lovat. In subsequent years he added the rentals of Pait and Killilan forests extending his domain from east to west. Cutting right across the Highlands, Winans set about various projects across this immense stretch of ground with the same energy he applied to railroading across continents. In an effort to contain all the deer on the properties for which he paid very considerable rents, Winans conceived, and set about the building of what became the longest deer fence in Scotland. The placing of hundreds of tons of iron materials along the planned route must have been the type of problem often faced by Winans in his business activities.

Under his instructions the local yard over at Loch Carron, the most convenient sea loch on the west coast, built a boat suitable for carrying cargo. The result, a heavy thirty foot flat-bottom vessel called *The Flyer* had then to be delivered to Loch Monar. She sailed across Loch Carron for Attadale and, ignominiously pulled by horses up the steep track to Ben Dronaig, was launched on the fresh waters of Loch Calavie to sail east towards Pait. After another shortish land haul, *The Flyer* took to the waters of 'An Gead Lochs' in its last stage before the downhill drag to Loch Monar. The boat must have weighed several tons at least, and moving it through such rough country was a feat to be admired.

The Flyer in commission carried hundreds of tons of materials up Loch Monar for the east end of the 'Great Fence'. When complete, and construction took a couple of years, the iron standard and dropper fence of the finest and heaviest materials stretched from the sea at Killilan in Kintail to the waters of Loch Monar at Pait. The line taken and the form of construction incensed the shooting tenant who marched to the north, for the fence was built in such a way, with rises of ground, that it allowed the deer from the neighbouring ground of Attadale to jump into the Pait forest but not back out again. The Winans fence within a year therefore gave rise to the construction of a second and parallel deer fence by the irate shooting tenant of the deer deprived estate, namely a Baron Schroder. The two fences ran for many miles sometimes as much as a mile apart. The Baron's fence ran west from Loch Monar along the lower slopes of Meall Mhor to the end of Ben Dronaig at which point it ceased and could have been effective in containing deer to the Attadale ground. Perhaps, however his money ran out. The latter fence being of poorer material did not stand so well but both fences were there in my day, symbolic of a high peak in shooting tenantry expenditure.

The locals certainly did not complain of Winan's excesses. He employed up to 300 men and in his grandiose way paid the highest of wages. This inflationary practice so upset the surrounding landlords they perforce were obliged to remonstrate with him. With colonial brashness he replied "The English stopped the slave trade in America, I've helped to stop it in the Highlands". The locals exulted at the story of this snub. However Winans did not always display such egalitarian principles as the famous 'Pet lamb' law suit plainly indicated. A well recorded story of how a Kintail crofter took Winans to the High Court in Edinburgh and won his case bears witness. A degree of high-handedness over, for him, a relatively trivial matter.

Winans was an excellent shot but his main interest lay not in stalking as we know it, but in deer driving. Using the natural defiles, sometimes with stone walls constructed on the ridges so as to deflect and concentrate the oncoming deer, he squatted behind a vantage stone as forty or fifty ghillies drove the animals towards his rifle. On other occasions he would walk behind the beaters as the deer were driven to a sheer cliff face or some impossible spot from which they were forced to break back. Winans showed his great markmanship and took the fleeing deer on the run. To gather his retinue of ghillies and beaters for a big drive the American used a system of signalling flags flown from various sighting points. One was still to be seen at Braulin in my day.

John Ross, the old Lovat Estate keeper at Ardchuilk, though he knew better than to say too much, was disgusted by what he considered to be the excesses of Winan's approach to deer killing. Ross preferred the skill of a conventional stalk and often remarked on suitably discreet occasions "The King of Sports is now likely to be spoilt for ever, just to gratify the selfish taste of an eccentric American".

Winans, perhaps with reason, did not trust the Highlander too far and at the start of one season ordered a count of all the deer on the lands which he tenanted. No small task, the keepers found themselves busy for a few days before Winans called them in to report. On receiving the numbers the American was both displeased and suspicious. He forthwith despatched a number of independent ghillies to do a double check. The second batch of results showed the keepers counts to be woefully low, hopelessly inaccurate in fact. Only one keeper had a count which tallied almost exactly with that of the ghillie. Incensed, Winans summoned all his keepers and issued the sternest reprimand. Probably he thought a venison trade of a rather different complexion was operating behind his back. The good honest keeper he warmly congratulated. Naturally in a get together afterwards the despondent stalkers, some of whom had feared for their jobs, enquired as to how the blue eyed boy achieved success in this difficult task. "Och" he shrugged, "why would I go to the bother that you chaps made of it. I took a walk up the glen that evening, spied every beast I could see, and then just doubled it". Though Winan's reign over this extensive area of the Highlands had a considerable effect upon local society during the 1870's and 80's, he is barely remembered today.

The end of Winan's boat came at the hands of another Pait Laird, Colonel Wills the tobacco magnate and owner of Killilan. Motor boats came into being and Wills put the first powered launch on to Monar. She was the *Seal* to be followed by a fine boat *Curlew* which ran for thirty years, and the teak constructed *Dunlin*.

A silting problem developed, shallowing the corran at the west end of Monar probably worsened by the altering of the burn mouth at Pait. To circumvent this major snag Wills hit on the idea of building a boathouse to the east of this obstruction. An elaborate boathouse and pier complete with rail tracks, boat cradles and winches for hauling out and maintainence was built a mile east of Pait on the shore of the main loch. Heavy materials for the new boat shed were loaded aboard Winan's old *Flyer* and the *Curlew* set off towing her up the loch. It became stormy. In the nature of fresh water expanses, steep waves developed rapidly. The tow boat dug in her nose and rose sluggishly. Suddenly without any warning the overladen craft failed to rise from a heavy wave. Winan's *Flyer* plunged to the bottom, nearly taking the *Curlew* down as well before the frightened men could cut her free to prevent a total disaster.

Winan's fence? Old Kenny MacKay coming to Pait, after the first world war as Stalker for a new Pait owner Colonel Haig, cut and transported lengths of the rich man's folly back to the lochside to enclose the fertile little hay fields, once the domain of Hamish Dhu.

By comparison the affairs of Monar Estate were a good deal less flamboyant. Purchased from Mr Holmes about 1890 by John Stirling, the deer forest formed an integral part of the stalking and farming activities of these fine estates. Best known and favourite of the Monar keepers was Tommy Fleming. A native of Loch Morar in Knoydart, old Tommy came to Strathmore as a boy in 1912 only to be called up as an engineer in the First War. In action he received a bullet wound in the leg, but like a fit hill man he quickly recovered, returning to Strathmore after hostilities ceased. He became head keeper to the Stirlings, moving to Monar in 1920. A clever and 'knacky' man Tommy could turn his hand to many skills. Before long his ingenuity put electricity into his keeper's home. The burn rumbling past the rear of his house was harnessed using a waterwheel taken from the old Struy mill, to run a generator and power the lighting of the house. Major Stirling, son of the original owner, arrived shortly after the big switch on. "By jove Tommy, you turned Monar into a town". Unfortunately the flow of power fluctuated in proportion to the flow of the burn. It was difficult to balance the system, and electric light bulbs suffered accordingly. Nevertheless few keepers showed such enterprise in 1920. Much to his wife Katie's delight, Tommy arranged the system to power a little engine for turning a butter churn.

Nor did Fleming's knack stop at electrical gadgetry. He made his own shoes and once a pair for his wife. From the Monar tweed lengths he turned out his own suits. Along with his enthusiasm for working the croft land at Monar in strict rotation, Tommy Fleming demonstrated the typical independence and self reliance necessary for living the isolated life.

However the burn at the back of the Monar keeper's house did more than power the electrical effects. With a wide catchment area and running through a narrow gorge above the home it had a propensity for flash flooding. Now it was necessary in those days, if single to employ a housekeeper, particularly if one's elevated station in life were that of Estate keeper. Fleming when first in Monar had a housekeeper for a short time. As duties required she rose first, and coming down one morning after a night of torrential rain, found to her horror downstairs flooded knee–deep in water which had burst open the back door. Most alarmed she called piteously upstairs to Tommy, who like most glen folk judged the suitability of the weather to the hour of rising, and was not unduly sharp on an inclement morning. "Oh, Mr Fleming, Mr Fleming the house is deeply flooded and more water is pouring in at the back door". Fleming, with a concern that certainly did not extend to rising from his blankets, called down, "Well woman open the front door and let it out".

A master story teller, Tommy would often, when about to embark on a fanciful yarn, presage it with the comment, "Well boys it's the truth I'm telling you this time". His stories were elastic, or variable to suit whatever time or occasion demanded. Invariably interjected as he thought necessary, should the listeners exhibit signs of listlessness, would come the phrase "Well now to make a story short". To the initiated this signalled a doubling of the yarn.

Fleming excelled as a stalker who knew each inch and mood of the ground he keepered. His retirement came in 1956 when sadly his eyesight began to fail. In addition to his prowess on the hill, Tommy's adroit handling of the gentry who came about at the season time, many thought his chief skill. Always patient, almost phlegmatic in dealing with their whims Tommy kept a military smartness about the tweed plus fours and in his bearing. Driving deer one day in Strath Mhuilich above Strathmore, the party paused for lunch on a knoll overlooking the cold dark lochan, from which the north running glen took its name. The Gentleman of the day, a trifle flustered and self conscious for he had hit nothing that morning, at least made a great display of emptying his rifle of ammunition before joining the luncheon. Thinking the weapon unloaded he pulled the trigger. The bullet blew a hole in the ground about a foot from where Fleming sat eating and contemplating the view. Tommy looked up unconcerned but with deliberation gathered his piece. "Ah well I think I'll move to a safer seat" and refused to make further comment.

Monar always had much coming and going of people, supplies

and the products from the hills. Katie Fleming, in common with other woman in the Highlands at the time, worked hard, milking the cows, cheese and butter making, in addition to rearing a family, feeding hungry men and visitors and attending to the general household work. Most mothers in the glen after the first war had their confinements away down country. However Mrs Fleming's second child, a daughter Jessie, was born in Monar due to a premature arrival. At the Strathmore end of the loch the last children born in the wilds would have been the MacRaes. Duncan MacRae and later his son Alister were shepherds in Strathmore after the turn of the century. Another Strathmore shepherd held the record for recent times with ten children. Jimmy Burns and his wife Margaret settled in the glen after the second world war, and added to their family each year. On one of the occasions of this annual event all was not well with Mrs Burns. November ice bound the loch, the boats lay frozen and the urgent message that his wife's condition required medical assistance was brought over to Pait by the worried shepherd. Iain MacKay in a marathon run, having broken through the ice at the corran, made Monar in 40 minutes. To great excitement, for rescue trips in those days were not commonplace, a summoned helicopter landed the local doctor behind Strathmore house. Fortunately all went well.

Not so happy were events in some of the houses west of us which in many ways were more isolated. They were all abandoned by the end of last century. At the foot of Corrie Ghraigh-Fhear on the pleasant south slopes of Meall Mhor stand the gaunt gables of a remote croft house, once the home of a long forgotten family who acted as deer watchers on the east boundary of Attadale estate. Beside the home is a large single boulder. In the adventurous fashion of children, a three year old boy of the family climbed the stone only to fall and break his neck. Grief stricken the father made a tiny coffin in which he carried the remains west to Clach an Duich. Teenie MacKay's grandmother, a child at this time in Corrie nan Each recalls the troubled father placing the box at the door of their house as he went in for refreshment. Like children they peeped inside. This happened over 120 years ago. Prior to this time infant deaths in the glens went unrecorded.

Deaths through accident were not common but Corrie nan Each west of Pait saw some little share of history. It may surprise many to know that the campaigning General Monck, Cromwell's henchman, passed through this Corrie. In 1654 an expedition designed and mounted to overawe the northern clans because of their attachment to the principle of a monarchy, set out from Perth that Spring. Commanded by Monck, the force of horse and foot included his own regiment, the now famous Coldstream Guards. This was a daring undertaking for those days. Apart from meeting hostility, the countryside could not support his men, thus with supply lines liable to become long and arduous, most food and equipment required to be carried by a large baggage section of heavy horse. From Perth they marched by Aberfeldy to Kingussie, thence along Loch Lagganside and over to Loch Lochy en route into Kintail. Reaching the sea at Loch Duich they swung inland east up Glen Elchaig over the Bealach and into Corrie nan Each. They marched easily downhill until reaching the swampy ground near Allt an Loin Fhoidha just south of Lochan Gobhlach and at the foot of Corrie nan Each. In his dispatch this camp is called 'Glen Teuch' which suggests the name was then probably Gleann t-Each, glen of the horses. His own words as follows '25th July, I came to Glenteuch in the Shields of Kintail, the night was very tempestuous and blew down most of the tents. In all this march we saw only two woman of the inhabitants and one man. 30th July. The Army marched from Glenteuch to Brouling. The way for near 5 miles so boggie that about 100 baggage horses were left behind and many other horses bogg'd or tir'd'. Never again any horsemen (much less an army) were observed to march that way.

More distantly in the fifteen hundreds there is record of a band of Strathglass Chisholms, who undertook a cattle reiving expedition into the Kintail lands of Clan MacRae. Returning home in rather a hurry after an abortive mission, the Chisholms were cut off by the MacRaes who with their better knowledge of the terrain reached the slopes of Ben Cruachan and stood waiting above the route taken by the unlucky reivers. In a fierce hand fight the Chisholms were killed to a man and their bodies dumped in a pool easily seen today at the east end of Corrie nan Each. This story and the identification of the swampy pool has been handed down locally by word of mouth.

Has the climate deteriorated from those days of higher population density in these glens? Have annual rainfall levels increased? Iain MacKay's great grandmother as a girl in Corrie nan Each told that the burn below their little 'but and ben' was easy to jump across. Today it is a wide rock strewn water course showing much damage to the greens which flank its banks. A good many shepherds in their day went barefoot to the hill during the summer, but any who did wear shoes had the task at the end of the day of scraping off the honey which gathered on the leather, so numerous were wild bees working on the rough grazings. Could a changing weather pattern influence the people in deciding to leave, or were the reasons purely of changing economic times? Old Kenny firmly maintained the summers were now wetter and hay-making more difficult, but then he was getting older and hadn't the same enthusiasm for the work. Records don't add up the attitudes of the people.

Many who did leave went on to make a way for themselves in spite of their humble background. I well recall a car sitting at Monar Pier one day when down for the mails. A fine looking gentleman stepped down to the boat and asked me courteously if I would run him a little way up the loch so that he might view the far away hills. The day, bright though it was, brought the hills in close, turning them darker blue, a sure hint to those who live by such signs of rain by the evening. During our conversation it transpired that the man now in his 70's had been born about six miles west of Pait at Maol Bhuidhe. The outline of the hills that stood above the ruins of his old home were clear and sharp. Renwick to name, his father had come to Maol Bhuidhe as a shepherd at the turn of the century. He spoke quietly of happy days when he belonged to a family of twelve, all born in that now desolate uninhabited area. As a young boy he climbed the hill to which he now gazed so intently. "I looked down to Monar from that top" he said pointing it out to me "and thought I was seeing all the big outside world". Pausing, he continued slowly and with some difficulty "I have never been back and will not be seeing these hills again". I judged it my place to leave him with his thoughts and said nothing. After a time I swung the boat, he didn't look back. Later I learned this fine Highlander was Moderator of The Free Church of Scotland. I read of his death after some little time with regret but understanding.

THE LAST STALK

S talking in its heyday might with justification have been called 'The King of Sports'. The nobility of the stag, his strength, fleetness, alertness and bold outlook has appealed down the centuries. Nor were the hunters any less needful of hardiness, fitness and cunning. The act of killing a fine stag as he stands upright, surveying his domain, forms a bridge to man's distant past and is not just idle fetish. Sad to say as I write the sport is now degraded in some areas to little better than a fairground shooting gallery for wealthy and often infirm foreigners. Enclosed deer shoots have been set up with platform hides. A drive up and shoot attitude is developing. How little do these 'sportsmen' seem to appreciate that a large measure of the stag's nobility stems from his freedom. The wind and weather are his boundaries. Head and antler held high, he scents the breeze in some high corrie, proudly he steps in the assurance of superiority in his domain of mountain and glen. Only men of like calibre should rightly kill such magnificence. Behind wires the stag is reduced to the degeneracy of his ignoble captors. captors.

A narrow pass just short of 3,000' lies hidden on the west ridge of

Sgur na Conbhaire. Steep rocks lead into a gloomy defile with a single track much used by sheep and deer as they travelled across from Toll A' Chaorachain to Pollach an Gorm according to shifts in the weather. Making through this pass on a day in early May when a chill wind funnelled into my face and sunlight of horizon stretching brilliance caught my world, I surprised a Golden Eagle which had been sunning itself, wings outstretched on a ledge below me. Unhurriedly it sailed off. Only one flap of a five foot wing spread. I looked down from above, noting so clearly how the fingered feathers at the wing tips curled up, appearing to move in the manner of an aircraft. The sun played on its dark reddish brown back. I likened the dark and light mixture to the colour of a Highland cow. It swept down five hundred feet before catching an up current and climbed with a soaring power unequalled in all nature, out over the distant shoulder of Beinn Tharsuinn. In many encounters with the eagle, perhaps this ranked as the finest I have had. How fittingly the King of Birds should impress me with its nobility above the heights of the Bowman's pass as we called the crossing between Toll a Chaorachain and Pollan Buidhe. This intriguing gash across the ridge had seen royalty on another day.

King James VI of Scotland was entertained by Lord Colin, First Earl of Seaforth. He and other noblemen followed the chase over the hills and passes that we inherited to shepherd. For weapons they used bows and arrows and long spears, killing the deer as they leapt through the narrow passage, and the memory of their sport has come down to us in its name, the Bowman's pass. In Brahan Castle there is a fine painting depicting the King of Scotland in Kintail - perhaps this was the occasion the artist recalled. At a point near the entrance to the Bowman's Pass there is a natural stone chair which we knew as Seachair Thomhais (Thomas' chair). The name recalls one Thomas of Fairburn, long forgotten, who used to sit at this commanding spot when viewing the chase.

The Royal deer drive was still an occasion in the 1860's. One such drive took place on the 5th October, 1867. Marching the north ridges of Strathmore lay the Achnashellach forest and a Mr Tennant of Leeds organised a drive for the entertainment of the Prince of Wales, later King Edward VII. The Prince arrived from Dunrobin Castle accompanied by the Duke of Sutherland. All the sportsmen from the surrounding shooting lodges were invited to meet the Prince and they enthusiastically responded to the call. Mr Holmes of Monar, with his stalker William Collie walked out through Strathmore arriving in ample time.

They were arranged on a bare ridge along with other sportsmen, all under the charge of a Head Stalker. Though badly placed to get a shot Holmes and Collie had a good position of surveillance. The Prince in the best location could be seen on the opposite side of the corrie. The wind direction was less easily arranged and perversely blew from the wrong bearing. Scores of beaters under the organisation and leadership of Simon, Lord Lovat, try as they might, could not force the deer towards the stance where His Royal Highness waited with keen anticipation, for on the move were some 120 fine stags. After milling and wheeling away from the Royal Entourage the ghillies could do nothing more than let the stampeding herd scatter in various directions.

Holmes, under the direction of the wily Collie moved to a better place and stood below a perpendicular crag, upright with his back to the rock wall. After a lapse of a few minutes they heard the clear sound of galloping hooves and heavy blowing animals rushing towards them along a beaten sheep track which passed close to the wall. The hunted stags dashed up so fast and close that the ambushing pair could have touched them. Holmes shot the first one without even raising the rifle to his shoulder and then a second at a distance of ten yards. It was not late of the day. Collie bled the two animals whilst Holmes carved his initials on their horns. Although they were only ordinary heads Mr Holmes wrote to his host for the day saying he would much like the trophies as a memento of the day. For some reason the letter did not draw a reply. So ended the Prince's deer drive, hardly a success, his Royal Highness had not fired a shot.

After 1860 under careful management and well watched, the deer populations expanded rapidly. The deer left their natural woodland habitat and took to the open hills. About this time Barbara MacLean, Iain MacKay's grandmother from Corrie nan Each, now married at Benula recalled seeing deer driven from Glen Cannich over to Strathfarrar. The hill would be moving with deer, "like a sheep gathering", she said "with dozens of men lined out on hillside and ridge". Prior to this Glen Strathfarrar held few deer and those seen confined themselves to the great Caledonian woodlands. Stalking as now practised on the open hills with much crawling and spying was not known. Thomas Fraser, Lord Lovat, a man of over six feet, didn't crawl on all fours or bend a knee when after the deer. In the woodland perhaps it did not become necessary to do so. Lord Thomas, it seems, was a brilliant natural shot. He would raise his elbow level to his shoulder, never bending his head or moving any part of his body except the finger that pulled the trigger. A clean kill invariably followed. As the deer herds moved to the bare hills, Lovat and his keeper John Ross, followed them. Ross, a protagonist of the new style of stalking would lead out the Laird in cautious exemplary fashion with great attention to the use of cover and dead ground. When they got near to a stag but not close enough for a shot, John would get very excited and instinctively drop to the ground to start crawling towards the animal on all fours, hoping no doubt that the Laird would follow his example. No, the giant Thomas would remain straight as a pole and keep on walking. In a minute or two he would bend down suddenly and catching Ross by the collar pull him to his feet. Both would move on until the stag showed signs of beating a retreat when if within two hundred yards Lovat raised rifle to shoulder and the stag without fail would drop, taken through the neck.

Not all sportsmen possessed such a gift. Old Tommy Fleming once, for a whole week, put up with a pleasant but trigger-happy guest of the Stirling family. The man was a buffoon and the hills were cleared of deer by his antics and wild shooting. On his last day Tommy stalked the man into a fine stag. Bang, the animal, surprised but unhurt, made off. However Fleming with incredulity saw a stag fall dead about a quarter of a mile away. They raced over, the guest danced delightedly about the kill and with some well chosen words of caution gave Tommy a ten pound note. In Fleming's eyes this went a long way to repay the damage of the week and allowed the guest to expand a little at the dinner table.

William Collie, the Monar keeper of the late eighteen hundreds, had to contend with a gentleman of similar inability. Mr Graham, K.C. of London, a charming gentleman, arrived at Monar very proud of his expensive new rifle. "The latest and finest model on the market", he expansively informed the sceptical Collie, who in his native way was as shrewd a judge of men as of deer. After a number of abortive days, Collie and the by now concerned and considerably deflated Graham were climbing Creag na h-Iolaire, a hill on the north side of the loch near Monar, when the wind abruptly changed right round and blew from behind them. This abrupt change of weather caused a mob of stags standing above them to rush to the top of the hill. The gentleman became very excited and fired round after round at the deer now fleeing madly over the skyline. None of the bullets took effect. Collie more than slightly exasperated, decided upon a sarcastic punishment.

"Mr Graham" he began with an air of gravity, "I feel much

concerned about your bullets. You may have killed Alister Mhor of Pait. Your rifle certainly pointed in the direction of his house and although it is three or four miles away, we are fully 2,000 feet above it, and I believe it is possible from your fine rifle for a bullet to carry that distance from an altitude of 2,000 feet". With a start of alarm Graham said hurriedly "Oh my God, it is possible, let us get to the summit of the hill and have a look at the house". Collie, fit and active, made a fast pace to the top with a show of mock concern. Mr Graham toiled behind, panting through a mixture of apprehension and unaccustomed exertion.

From the top, even with the naked eye, a big black object, lying in Sandy's potato patch could be clearly discerned. Mr Graham saw it. "God bless me" he exclaimed. Collie through the glass perceived that the black spot was in fact Alister Mhor, and without a word handed the glass to Mr Graham. Clapping the telescope hurriedly to his eye "upon my word" he groaned as the distant spot materialised into the stooping figure of big Alister apparently wounded. The poor man sat down and began to mop a now very pale face. Even Collie himself became uneasy, thinking of the adage, 'many a true word is said in jest'. Graham realising his position as a gentleman now took command. "I suggest we go over at once Collie, in case medical assistance is required" he boomed. Collie spied on, totally ignoring the trembling man. After some five minutes Big Alister got up from his potatoes and walked into the house apparently none the worse for the stray bullets.

September was the stalking month in Strathmore. The Stirling family in the 1950's owned both Monar and Pait Estates. Their attitude towards the sport and management of deer forests being the same as those of a hundred years ago, it was a pleasure to help in the long days that were often involved. The stags would not have begun the rut and would still be on the highest ground in prime condition. Sir John and Lady Stirling generally stayed at Pait Lodge with a cook and maid to help run the house. On a typical day in mid September, preparations began at about 9 a.m. Stirling and Old Kenny his stalker, held conference in the yard deciding the order of the day. Yesterday's activities and the area stalked over, today's weather and wind direction, stags that had been recently seen and where they were grazing, who was available to help, how many rifles might wish to go out – all such factors were taken into account before a plan was formed between them and the party moved off.

The stags selected the best of the grazings and therefore the great broad face of An Riabhachan often became the scene for the day. My place fell either to act as ghillie for Old Kenny the stalker or sometimes pony man to follow on behind the shooting party handling a string of ponies. As a ghillie with the party my duty was to carry the sportsman's rifle, walking in single file behind the stalker who led the way with the shooter who came next. Each vantage spot required us to spy, ascertaining where the deer might be placed that day. The party moved on cautiously. Once well out on the ground little talking took place and then only in whispers. The success of the day was dependent upon the skill of the stalker in charge. One followed his movements, crawling when required, walking up a burn, crouching, lying, be the ground wet or dry, anything that allowed the party to get close to the stag that he had selected for the kill.

Often the hinds, always the more wary, would come between us and the quarry, perhaps a fine stag lying along on a high ridge. In the middle of the day a herd often lay cudding contentedly full until their afternoon grazing period roused them to move on. The stalkers ability to get around such a problem made or marred the day. However once the rut commenced early in October when the stags 'broke out' amongst the females, the nature of the problem altered. A stag with his selected group of females often took a rest during the day, not without good reason. The watchful champion generally selected an eminence from which he could control his harem and scan the surrounding hillsides for rival stags roaming free, keenly ready to move in and challenge. At intervals, night and day, he would rise and, head stretched out, let out a long far-carrying roar, especially in response to another stag's bellowing from across the corries.

A powerful stag from time to time would dive down and quite roughly herd his group of hinds together, rather as a collie dog will circle a cut of sheep. The females, respectful of his lordship but not evidently loyal, might be cut out and driven off by a successful challenger. A stag driving captured hinds worked them away in great haste in the style of a sheepdog darting from side to side. The whole interplay of the stag's raw power contrasting with the fickle submissiveness of the females gave a fascinating example of crude natural selection.

Old Kenny, a patient and gifted stalker, knew many facets of deer behaviour and would often anticipate the likely movement of herd or stag to the advantage of a shot. Wind direction could be a constant problem for the stalker and an intimate grasp of air movements a mile away up corrie or over ridge, came with intelligent observation of the weather. The deer being more scent than sight sensitive would be off at the least puff of wind which carried man to their nostrils. On a damp day the human scent carries sometimes for half a mile or more. Any slight movement the hinds in particular would detect over a considerable distance, but not generally at much beyond a mile. Provided one stayed perfectly still, (although in full vision) even as close to deer as a few hundred yards, one could escape detection. It seemed to me that deer, though sharp on movement, had less ability in colour differentiation. Nevertheless we always wore tweeds of background blending shades, and no ghillie went without a "fore and aft" bonnet.

The ponyman missed the thrill of a stalk and the job often involved sheltering cold and wet behind a stone, wishing the stalking party would get a quick shot. From a stance down the glen the ponyman with a quick eye could watch a stalk, sometimes seeing more of the overall activity than the shooting party, but telling where a day had gone wrong would be spoken of only with discretion.

We generally took out three ponies. The stalking party left for the hill. The ponyman busied himself giving a last small feed to the horses before securing the deer saddles. One pony was led and the other two, tied in file, followed obediently. Woe betide the ponyman who came up too soon to his allotted stance, or worse, went along the path too far. The stalk could be ruined by the infuriating sight of a ponyman, blissfully unaware, tramping boldly up the path, a string of ponies clip-clopping behind him. Waiting at a shelter stone I would make myself as comfortable as possible, loosen saddle girths and tether the ponies so there was no chance of their bolting off home. For extra safety I often tied a line with a lead to my foot, and lay waiting, sunning myself if the day blew fair. On hearing the long awaited shot, and my ear could detect the subtle difference between the hollow sound indicating a miss, and the dead or heavy explosion indicating a hit, the ponies' heads jerked up. They knew the day was on the move. Relieved and glad of action the good ponyman did not, however jump up and show himself, but waited. More shots might follow if the day had gone wrong with a miss or a wounding, or if really well another stag might yet run in or move into range. It was essential to remain out of sight, watching and searching. In dry weather the keeper often lit a small fire to signal up the ponies or stood and waved hat or hankie. The lowly ponyman had to spot the order and respond. Saddle girths tightened, he used his judgement to get his team as close to the kill as the ground would allow.

The scene of the kill was always interesting. The stag, if a good one was generally dark in colour. A neck full and bold betokened strength. His head or antler brought comment and admiration. A dark thick horn wide spread between well-developed three-point cups at the antler ends was a rare trophy. Long brow points, evenly curved, set off the intervening points to make a truly fine head — perhaps a 'royal' twelve pointer. The Monarch would be stretched out ready for loading. The ghillie had the task of gralloching the beast. Anyone of gentle stomach would not relish the smell of warm blood or the gut content of the poca buidhe. To ensure a good drain of blood from the carcase the stag's throat would have been cut. The stomach opened from the penis forward to the chest bone and backwards to the tail allowed the intestines to be wrenched out. A good heavy stag might weigh 17 — 20 stone, so to load it on a pony required both knack and strength, apart from the first essential off getting the pony to stand still. Though the ponies had no fear of blood or carcase they might perhaps object to the weight, and annoyingly would side-step at the critical moment and leave a cursing ghillie with the deer in a heap at his feet. It was easier to load a stag with two men, one pulling up the head whilst the other, less senior of course, put his shoulder under the stag's tail, often to get a shower of blood down his neck if he wasn't careful. Long leather straps secured the unwieldly burden round antler and rump, and the pony stood ready for the journey down to the larder.

Dandy, the hardiest of ponies, small though he was, when once loaded would toddle off home at a fine speed, stag swaying high above his body on the heavy deer saddle. This pony had the sense to stop immediately, should the stag start to slip round his saddle. We often let him go on ahead if there happened to be another pony to load. He would not make any mistake either of path or load and would be waiting at the larder or more likely trying to get into the stable for his feed when we caught up with him. With commendable manners, we were amused that he never dirtied the path, but would step up the hill a little and then carry on.

Not so clever was the mare Shean that came up from Fairburn to work with Dandy. Coming off the grass from the home farms at the start of the season, she was fat and out of condition. The party were stalking on Ben Bheag that day and shot a good stag. Duly loaded, a certain ghillie set off for home with the puffing Shean meekly following. Whom we should deem the more stupid it is hard to know but the ghillie led the loaded pony into a bottomless bog. No wise pony would have ventured into such a morass. The wily Dandy we often saw try an area with his front foot very gingerly before deciding what to do. Not Shean, all four feet straight in, down she went almost out of sight. Iain and I shortly came on the scene. The panicking ghillie had managed to pull off the stag and, arms black to the shoulder in peat struggled with the saddle. The mare floundered about snorting in terror. We brought Dandy into action, tied all our drag ropes together and attempted to pull her free. The stupid brute wouldn't try to help herself. Being late, plans were laid for the morning. "If she is still alive,' Iain expressed the thought in our minds.

I rose early the next morning and went up to the great bog which lay about two miles from Pait below the shoulder of Ben Bheag. To my surprise the pony was still alive and had moved a little through the night, but still belly deep she looked cold and dejected. Methods were adopted which would have been frowned upon by the gentleman, with us the previous evening. Taking an end of the rope I carried, without mercy I lashed the miserable creature with all my force. Squealing, snorting and trying to kick me, she lunged forward as I laid on the lash. By this brutal method I got her towards firmer ground. Feeling her feet she crawled out on numb peat caked legs. I caught up the exhausted and trembling pony and led her down the path towards Pait. Nicely timed as a boatload of helpers with spades, ropes, pulley block and lifting equipment grunted up the path towards me. It must have pleased them when the pony hove into view none the worse except for a few weals across her hefty quarters to which I chose not to refer.

My employers, Sir John and Lady Stirling, had very different but distinctive characters and on their trips to Monar whether for fishing in June, stalking in September, or hind shooting in December I was often in their company. Sir John had been struggling with an arthritic hip for many years but whatever pain he suffered did not dull his spirit or his sharp twinkling eye. From the back of Strathmore cottage I watched him and keeper Allan Fleming climb out along the ridge of Creag na Gaoidhe. The October day blew a keen strong wind filled with cutting sleet showers; I watched them come out on the top of the exposed ridge 2,000' above me and turn westward. Allan leading with the rifle, Sir John, a stooping huddled figure wearing an old army gas cape which billowed out in the gale, limped along behind, his inevitable stick very much an essential that day. One false step I knew only too well meant a slide of a thousand feet. The deer sheltered from the severity of the day, and as often is the way were more easily approached in adverse weather. They got a stag at the west end of the ridge and on hearing the shot I went up with the pony. Sir John, blue with cold, beamed cheerfully but I could see he was sore, though as ever he said nothing. About seventy years of age at this time, how I admired his spirit and toughness.

Lady Stirling, "Katie" was, by any standards, a formidable woman. Daughter of that old Highland family, the MacKenzie's of Conon and Gairloch, she combined iron will and constitution, brooking no interference with any plans she might make. Her family she saw to it did not overeat, indeed Lady Stirling was a most frugal housekeeper and believed in plain living. Tall, erect, steel grey hair and commanding, one eccentric aspect of her character came to me as a surprise when early in my acquaintance with them both, we were all sailing west to Strathmore in *An Gead* on a fresh squally day. Lady Stirling in her Monar tweed knicker suit sat in the bow for a little shelter as the boat danced on the lively waves. To my amazement, though I tried not to show it, she took out a tobacco pouch and little clay pipe. Soon the fragrance of tobacco smoke blew down to me as Lady Stirling puffed away. For some reason she put the pipe down and it vanished under the bottom boards with the heaving of the boat. Windy day or not Sir John and I had to lift all the floor boards until we came on the pipe lying in the sloshing bilges. By way of explanation Lady Stirling told me she had picked up the pipe habit from the fishermen at Gairloch when she was a young girl.

Yes, her Ladyship could make things tick, she ran the Lodge affairs with firmness making sure there was no wastage, and, perhaps a more difficult task, attempting to protect the young lassies who came to help as maids from marauding ghillies or itinerant shepherds. Behind all her fierceness Lady Stirling showed a great kindness and sincerity for the welfare of many less fortunate people.

My interest in red deer and stalking matters soon became equal to that of my attention to the subtle art of sheep husbandry. No day passed without my seeing deer near or far. Due to the constant exercise of changing focal distance my eyesight became quick and sharp. Having worn reading glasses during my army service I quickly found my vision so improved that I have not required them since. The deer often blend so effectively with their background, only the keenest eye would pick them out.

During the summers, humid midgey weather drove the herds of hinds and calves to the ridges which the breeze kept pest free. However I remember on one occasion surveying the wide country about me from the top of Meall Mhor several miles to the west. The green ground around Loch Cruishie caught a yellow spoke of sunshine. It lit the bonnie lochan's white sandy edges, down to which the lines of a past generation's cultivations ran in green memories of a lifestyle long gone.

The shallow lochan sparkled in the sunlight and grazing idly across, I caught sight of objects like specks on a silvery lens. It was a hot sultry day, with blinks of sun and warm puffs of wind. I spied down, and saw the dots through my telescope were deer, some fifty or sixty, hinds and calves, standing motionless like cattle, belly deep in the water. I had not before seen deer adopt this method of protecting themselves from the scourge of cleg, fly and midge. Although apparently incongruous it betokened the safety that these animals must have felt, secure in their remoteness from human disturbance.

In these mid-summer doldrums the stags run together in large groups, keeping themselves separate from the domestic scene for which the previous October made them responsible. Commandeering the best grazing areas they live peaceably together, rapidly assuming a fine July condition. The warble fly grubs which erupt through the skin along their backs in April have ceased to irritate them. Antlers in velvet and full of blood now mature and commence peeling to reveal their set of points. Hot July days see these bands of stags search out the dwindling snow fields, there to lie or roll as they strive to cool themselves. In August the mature stags will have rubbed clean the velvet from their antlers and be carrying a spread which reaches its best dimensions between eight and twelve years of age. Strathmore, because of the quality of its grazing, carried many stags with fine heads. Thick at the base with long brow points a good set would sweep wide with two more points to a well spread cup of three points giving twelve in all and a Royal stag. It was the policy of the Stirlings not to shoot the best heads, but rather to cull out poorer specimens. Strathmore Lodge had a most unusual collection of freak heads. Many were malformed in surprising ways, for what reason I was never clear. A 'Hummel' stag, that is one without any horns at all, would always be a heavy chap and a great prize to shoot.

The rutting season commences as the weather cools and the governing factor of daylight triggers male hormonal activities, generally during the last week of September or early October. The stag bands quickly broke up and the hind herds are invaded. A stag cuts out as many hinds as he can hold together in a group. A big stag might often control a harem of thirty hinds until tired or beaten. Wandering stags without hinds move freely about the hills constantly challenging a master for a share of his attendant females. At this time for a month or more the noise at our house from the roaring stags could be exceptional. They seemed to choose late evening to be at their most vociferous and the long drawn out bellowing would echo round the hills as stag answered stag.

Though much is made of the stag's ferocity I didn't see a great deal of actual fighting. A good stag would soon chase off an interloper who it seemed knew how close he could approach and still be able to escape with a quick sprint. However I did witness a splendid fight one day on the clean grassy slopes of Sgurr na Conbhaire east of the great fox cairn.

Late one October afternoon the sky cleared to the west over the cool darkening peak of Bidean a Choire Sheasgaich. The air had a tingle of oncoming frost and smelt of summer's dying beauty. A heavy stag held a cut of at least twenty hinds. In a restlessly savage manner he kept running at any unwise hind that moved out of quite a tight circle. He seemed to be attempting to drive the group before him. I was at a distance of quarter of a mile and in full view as I walked, dogs at heel, unhurriedly along the path. The stag was so preoccupied with his hinds, I sat down unnoticed on a boulder thinking to watch for a few minutes.

Almost at once, coming over the shoulder out of Toll a Chaorachain, I spotted a massive black stag. He came galloping down towards the party of hinds roaring as he ran. Immediately the first stag turned to face him standing his ground and bellowing back up. The hinds' heads turned to watch the oncoming beast. To my amazement without pausing for any preliminary sparring the stags flew at one another. Antlers lowered they met head on with a clash that clearly reached me. A fight engaged in deadly earnest. Not taking their heads apart each pushed and wheeled trying for the advantage of higher ground. They were obviously evenly matched in weight and head as they dug and spun rings on the ground. The grunting and gasping floated down, easily audible to me. The hinds grazed rapidly away from the battle taking no notice whatsoever.

Suddenly with a slither of flying mud one stag went down. Quick as a knife stab the top stag flashed a thrusting jab with his head into his adversary's side. The fallen victim rolled right over but leapt on his feet before a second belly thrust could be driven home. Down the hill he fled as the victor caught him heavily in the rump. They ran for at least 150 yards before the champion stopped and stood tongue out, head down, panting.

The vanquished rival did not pause but made down to the river. I could see he was hurt and bleeding. Stumbling down the bank he forded the water and made his way very slowly along the opposite bank. Meantime the other stag was climbing back to the hinds. It looked to me, though both stags' coats were mud besplattered, the fellow who had run in claimed the victory. The driving action of the stag which I had first noticed suggested he might have stolen this cut of hinds whilst the big chap's back was turned. If that was the case then he certainly paid a high price for his impudence.

It was my privilege to accompany Lady Stirling and old Kenny MacKay on their last stalk together. They chose a Saturday in late September. The MacKay boys and other ghillies being away for the weekend, I was invited to give a hand. The three of us met in the stable yard at Pait. Busying myself putting a deer saddle on the ever willing Dandy, I gave him an extra handful of bruised oats as the voices of Kenny and Lady Stirling making plans for the day carried into the dim stable. We were for the distant Riabhachan face I gathered from their conversation. Old Kenny led with a springy step that never left him to his very last days. Lady Stirling with myself leading Dandy followed him up the path winding southwards, hugging the morning bright waters of a roguish burn which had its source at nearly 4,000 feet in crystal cold springs called the Wells of Riabhachan.

The fresh, steady southwest wind which caught our faces was bracing and invigorating. Never did I feel more alive to the sunlight. It warmed face and arms as we climbed gradually out past the top hill park gate. Lightsome we stepped out to the sheiling greens of yesteryear, waiting yet long forgotten, sparkling with a canopy of cobweb-laced dew on their maturing unwanted grasses.   To our right lay the purple shoulder of Ben Bheag heavy in deep heather. A grouse rose noisily, its "go back, go back" reminding Kenny of his youthful keepering days and the once heavy bags of game taken from this ground after the turn of the century. We didn't speak. My eye followed the bird to a heather knoll where it landed with gliding flight.

The clip-clop of Dandy's feet on the stony path told a private world of our progress. Ahead, green and peaceful in the thin sharp light, lay the rolling ridges and corries of the great Riabhachan face. Turning left the path dipped down to the boisterous burn which spread over the black glistening slabs of a waterall in a curtain of sunlight patterned water.

A red, autumn berried Rowan grew with arching trunk from a dark crevice, in bold contrast to the clear sky. Flitting down to the burn to warn us of territorial rights came a grey wagtail. I was surprised to see the bird so late in the season. These water sprayed rocks I knew to be its nesting haunt, probably in a moss lined cranny beside the falls. The dainty bird sadly misnamed grey, showed its sulphur yellow chest and blue-grey wings as it flicked a long tail from each rock perch, calling with a tiny squeaking alarm note.

Stalker and sportswoman crossed the foaming waters by means of a flimsy wire swing bridge which dipped and swayed ten feet above the burn. I mounted and crossed with Dandy. A sharp pull up the bank brought us to a rumble of stones fashioned into a crude shelter. This low knoll commanded a panoramic spread of hill ground, majestic in sheer scale and dominance over human arrogance. It was a natural spying place and the ponyman's waiting stance. Old Kenny never hurried a day, and here we sat, our thick tweed suits making comfortable a damp cold seat. Emanating quiet enthusiasm, telescope steadied by stick, Kenny began a deliberate study of the position of various groups of deer spread before us over thousands of acres. Having carried up the rifle and supplies I took a quick spy and then, appetite edged by the vigorous day, ate half of my venison piece. Lady Stirling sat quietly, content in the beauty of the day. Kenny directed me to spy a towering eminence sweeping up from the Riabhachan wells and just as he indicated I picked out a useful stag lying quietly above a group of perhaps eight or ten hinds.

It approached midday, the warm sun now over the twin peaks of An Socach, which nodded in shimmering outline over the deep shaded bowl of the green corrie. To the rustle of twitching bracken and the faint tinkle of the endless burn I could have slept. Kenny stood up and moved off. I had brought extra rope and securely tied the cunning Dandy, with a bowline knot around his neck, to a rock that he could circle. In case he played at rolling, I took off his saddle, and gathering the rifle hastened after the eager pair.

We took a long steep runner, or small valleyed water course, far to the east of where our quarry lay. The climb turned wet and slippery. An ideal wind put our scent down and away from the deer, but tackety hill shoes clink on the stones, their sound carrying far, so I stepped lightly and with much care. To our left a large group of hinds was controlled by a fine aristocratic black stag. They did scent our wind and made off at a high stepping trot. What a magnificent brute he showed himself to be when moving, a great neck, black from rolling in the peat, a proud head and carriage which set off his wide sweeping antlers. Head held high he strode after his nervous hinds. I knew

Kenny would ignore him, for a stag of that calibre would leave his mark on the quality of the stock through his offspring. Fortunately, as the stalker rightly anticipated, the curve of the long face kept the fleeing deer out of sight of our intended group.

Finally, we reached the top of the steep moss–pooled runner and started to cross firm open ground. We headed diagonally west and gently upwards towards the shoulder on which we hoped the deer might still be resting. Kenny crouching as he walked, stopped and spied frequently, taking the utmost care to open and close his little two draw telescope without a sound. No word passed amongst us. We moved without a sound. Abruptly Kenny stopped, I knew he had sighted the deer we stalked. Hand out behind to warn us, the old man sank imperceptibly to his knees and spied again. Close behind him we acted likewise. Without a word Kenny began to crawl stealthily forward. The stalker suddenly lay face down. In line, down we went, bodies on the mossy wet ground. Every stone must now be used to keep us in dead ground from the dozing deer. Lady Stirling crawled as we did, her boots and thick socks just a foot from my face. I took care to keep the rifle dry, free from dirt yet certainly close to the ground.

Face to the ground I crawled on. After what seemed hundreds of yards we stopped. I knew the deer to be close, in fact such is the keenness of senses I could smell their acrid scent borne on the light steady breeze. Lady Stirling inched gradually forward and lay beside the stalker. Only a smallish boulder gave cover. I moved a little not daring to breathe. Kenny looked back and motioned the rifle. Taking an age to clear it from the case, I slid it forward. I looked up very slowly keeping my chin pressed to the ground. There on the knoll lay a fine stag not eighty yards away. He flicked an ear and cudded. In his full view I hugged the ground. The hinds below him were not in my sight. I lay flat and waited. The water soaked into my tweeds.

Turning my head to the side at a sound I saw Lady Stirling fumbling with her glasses case. The rifle lay ready before her. The stag instantly alert, stared fixedly down, he had caught the sound. Up he got, head on to us. Time turned to heart beats. He eased, turned his head up the hill. Magnificent; personification of the wild power of hill and glen, this handsome beast now broadside to us, outlined against ridge and sky, threw up his great head and roared a challenge. "Now" I heard. *Crack* . . In an exploding second his world of bright sun and pure freedom crashed to the ground. Dead; shot through the heart. Dead, before the echo of his brave call to life could float back from the far corrie wall.

We lay on without moving. Kenny spied again, then satisfied the stag lay dead, he stood up. "Well done, you got him." Lady Stirling was overcome. She thanked us both with some emotion, obviously feeling that deep strange mixture of lust and regret which comes at a great kill. I went over and bled the still twitching stag at the neck. His eyes stared out of memory.

I wiped the blooded knife on his proud dark mane. Out spread his red sacrifice on the cool green moss. Kenny sat down and took out his pipe. Lady Stirling moved to a boulder and they both looked away to the hills of the west, clear in the falling sun. Old Kenny knew retirement was at hand. The Lady knew changes were being forced on the glen. Neither spoke. After a while they talked a little of bygone days, the wistfulness of inner thoughts showed through their conversation.

However Lady Stirling not to be down hearted said "Come on Kenny we are both seventy. I'll race you to the top of Riabhachan". "Och you can be going if you like, I'll attend to the stag". The old lady set off with a will and climbed the remaining five hundred feet to the top. It would be for the last time, that she somehow knew. My job was clear, I ran down for Dandy who spotted me coming from a distance and pleased something was happening at last, gave me a call. We soon picked our way back up to where the veteran stalker waited, stag gralloched and ready to load. Dandy was such a handy pony and would have lain down to make our loading easy if he could. We put him alongside a bank. He stood rock firm and with a quick pull I slid the heavy stag onto the saddle. "You'd better wait till she comes down" instructed Kenny, then "what a woman!" he laughed, shaking his head. He set off down the hillside with the tricky job of leading a pony on steep ground, for the stag is always likely to fall forwards. I waited, watching with admiration old Kenny pick out the easiest way down below on the fading slopes.

By and by Lady Stirling joined me. The lowering sun spread the colour of life far away on the Applecross and Gairloch hills and she looked long and thoughtful, perhaps to scenes of her youth. I saw she was in tears. Around us in the purple light of evening there was a chill to the air. Gently I led the way down. She was very tired. We talked quietly of many things. Loch Monar lay dark, as we retraced our morning steps past the hill park gate. The lights of Pait Lodge held a welcome in their twinkle.

Old Kenny made down long before us, the stag lay at the larder door. Dandy I heard at his feed in the stable. Mrs MacKay had on a fine meal of her speciality, potted venison tongue. Kenny and I needed no encouragement. Later I took out a lantern to the larder and skinned the stag. Lady Stirling came down from the Lodge to look at the carcase and thanked me again. It was a still night, the light from the window of Strathmore reflected a zig zag line across the calm waters. I rowed home to where Betty with our children in bed waited to hear the events of the day.

IN WINTER'S GRIP

A portent of winter's steady approach, is the first faint scatter of snow dressing the high tops of Strathmore during late September. Lying perhaps for a couple of hours in the wan morning light there is still enough warmth in the southing sun to clear the peppered flecks by midday. The air feels different, man and beast sense the coming change and react according to age old instincts. A vague almost intangible feeling of unease pervades. The flow of migration heightens apprehension in those who remain. Winter's threat weighs on mind and action. A need to prepare is felt deep in the very bones of existence.

Sheep must be gathered in early October for their pre-winter waterproofing dip. It is our last visit of the season to the summer pathways of peak and corrie. The changing scene from the lofty ground tells its own story. Grey curtains of rain hang on the horizon. Chill and sweeping, they cut into dull brown hills pushing the clinging mists of summer heat before them. There is a sense of subduedness, for the impending struggle of living through the barren winter is at hand.

One day early in October I reached the narrow sheltered ridge of Bidean with such a feeling within my inner self. Away to the west the view was obscured by low folds of iron cloud driven by the Atlantic winds. Below me, the deer corrie lochan which sparkled only months ago with emerald light was that day, a menacing wave-flecked black surface. I felt an involuntary shudder as I looked down. The cold damp wind of a changing day blew with plucking fingers at shirt, neck and jacket. Not a day to stand. Mists began to race towards me, precursors of stinging rain which was now driving hard and relentlessly across the Attadale hills. Dropping down a few hundred feet from the crest I came to the bubbling spring which rises clear from a cleft of rock at the head of Coire nam Minn. Of artesian force, its flow remained unchecked summer or winter, mysterious welling bubbles lipped by green mosses. I paused for a mouthful. Its coldness stung my throat but its sweetness held the elixir of life for those who cared to drink and ask of the ageless hills their secret.

Stepping out to the flat curving plateau of Sgurr A' Chaorachain, now bleak with its crumbled shattered rocks, I was met by a rolling wall of shivering mist. To be certain of my bearings, I paused to note familiar objects in a world which suddenly changed its dimensions. I stood, the dogs beside me, winds ruffling their coats as they sniffed into the enveloping fog. Suddenly wheeling in, it seemed from nowhere, not ten yards from me, landed a large flock of Golden Plover. They alighted in unison with such grace of movement. Each bird as they touched the ground raised beautifully long slender wings to arch and meet tip to tip over a grey feathered back. Gone was the glossy black breast of courting spring days. Gone the golden back which had glinted in the summer sun. In place was a quiet winter dress, smart and storm proof, to carry them many weary leagues to some sun locked strand in Northern Africa. They perched to rest. I stood to watch, conscious of some inner calling. Buffeted by the swirling mists, they shuffled feet to balance in the crude gale's thrust. Without warning, after some minutes, again with a unity of purpose, they lifted their wings to display the downy white underneath. Then giving no signal they took silent flight, instantly to be swallowed by the rain filled vapours. In a second however their enigmatic whistle, an echo from summer ridges, caught my ear. They headed south. I made my way carefully first, for even known rocks become false friends, and then I prodded my way down anxious to get below the mist and into safe ground. The long-fingered hand of winter was clawing its way into the glen.

Outlying sheep had to be pulled in, often from the grounds of neighbouring estates. Dipped and painted with bright red keel they were easily recognised by their darker colour from the white coated ewes that had missed these gatherings. The lamb crop was drawn off and along with the cast ewes, or ewes too old to remain another season on the ground, we drove big flocks down to Monar and onto the transport waiting to take them to market. There was a close relationship between sheep prices and shepherds' wages at that time which is no longer the case when comparing with today's ratios. In due course I ran about 40 black faced breeding ewes of my own which allowed as the shepherds 'pack', were a perquisite of the job. Our own lambs went away with the estate sheep, and marked separately they became the first livestock I ever sold under my own name. Good fresh lambs they made 84 shillings a head in the Inverness auction mart. My wage at that time totalled eight pounds per week, less the various Government deductions. On that same day old Kenny sold three year-old wedders which being heavy and in good condition made £6.10.0. roughly equalling the shepherd's weekly wage per head. It took two lambs to make my pay. It now takes approximately five lambs to make up a shepherd's wage.

These autumn sales saw the culmination of a year's work and care. With any additional expenses incurred very likely to depress profits below an acceptable level, it became a time of financial worry for many estates. For us it held the delights of a sale's bustle with the added thrill of seeing stock go through the ring which had been the personal achievement of caring hard work. Days spent away at the sales meant somebody had to stay at home to tend the stock and invariably it was the women folk of either household. The two habitations were separated by boat and long paths, so it was a comfort to see that all was well when the first morning smoke drifted from respective chimneys.

After the MacKays retired and the boys left, we were alone for twelve months in Strathmore. Our neighbours were greatly missed in many ways. It is true to say that glen living depended for its practical existence on neighbourliness and the general communal help. Never could one wish for better neighbours than our friends at Pait. I ran Teenie and old Kenny down to Monar with the last of their flitting. Forty years of living in surroundings of complete familiarity is hard to put aside and I saw how keenly the parting was felt as we sailed down that afternoon. Soon we were to feel the difference. I had lambs to sell in Inverness and notice arrived that sheep belonging to Strathmore and the MacKays had been picked up away west at Ben Dronaig. The letter informed me that the strays lay awaiting collection out at fanks above Attadale. It suited to combine the two jobs.

Betty, children, myself and dogs left Strathmore after dinner on the following day in the *Spray* and sailed down to Monar. Fortunately, not a stormy day, I had persuaded Betty that she could easily manage the boat and the navigating. She looked a little doubtful but after gathering the mails at Monar pier, no objection was made and I watched them sail away west to round the point into the narrows. There was no way of knowing how they would get on. The children waved as the boat vanished from my sight.

Taking Nancy I set off for Inverness with a lift from the 'postie' down the pot holed miles to Struy. For one reason or another I stayed away over two nights and on the third day a plan was made for Sandy, the Fairburn shepherd to run me round to Attadale by Land Rover. A grand stategy, we would go across to the awaiting sheep and he could then take home to Fairburn any old creatures that might not stand the long drive back to Strathmore.

The day started in fine form considering the previous two nights. Could it have been the fumes of the vehicle? By midday we both succumbed, victims to a distinct and debilitating thirst. By unanimous vote we pulled into the Station bar on the platform at Achnashellach. Two trains and several hours were required to quench our affliction. By three o'clock, however it became abundantly clear even to us in the warm sanctity of a convivial hostelry that time and daylight were becoming distinctly limited. We sped off round to Attadale, twice circled the 'Big House' to correct our orientation, before, with masterly insight on the original purpose of the day, we chose the long rough pull up to the fanks at Ben Dronaig. Here we were met by a motley collection of wide eyed sheep, some of them Estate, some of old Kenny's and one or two, my own. Man and beast viewed each other with certain misgivings. With the enthusiasm of crusaders, in not many minutes Sandy and I divided the startled animals according to their category and condition. The old 'has beens' were bundled unceremoniously into the back of the Land Rover and with an unsteady but optimistic wave, in a shower of tyre spurted gravel, Sandy shot away. It was getting late. I looked dubiously out to the fast fading hills and thought of the twelve mile hike home. Nancy seemed keen and driving about fifty ewes we set off.

The route climbed, a faint track high above a steep slope that fell into a dangerous many waterfalled burn which led into the oppressively narrow pass of Bealach A Sgoltaidh. Adding to the gloom mist began to form with the darkness. We toiled upwards. I now depended on the leading sheep to stay on the track themselves and Nancy to hold any side wanderers. Black cliffs hung with foreboding over the ominous pass, adding evil to the darkness. Mist swirled up from the flats far below catching a rock, a crevice, a gully here and there, like the touch of white hands clawing with death's clamminess over grotesque faces. The wind mocked with a hollow laugh amongst bare stones. It grew chilly. I began to think ahead and wonder uneasily for the first time if the boat had reached Strathmore. On into the black chasm, now working by instinct, glad of my ability to see well in the dark, I hurried on. With great relief I felt the track under my feet turn downhill and dimly against the odd star which now broke through the gloom I could see the outline of my own hills of Strathmore.

The sheep weary and hungry after their imprisonment became increasingly difficult to handle. My aim had been to reach the stone fank halfway down Strathmore glen, but still some distance from this objective, I peered ahead, my eyes searching the cursed night for a glimpse of the lights of home. From a point I knew the house to be in view, no yellow dot of welcome could be seen. I became prey to increasingly wild conjecture. Finally with snapping resolve the sheep were abandoned. I ran, and ran hard, the remaining three or four miles down the rock strewn stumbling path to the gate of home. Not once did I see a light. My mind jumped alive with horrors. After all a woman and two small children had been left three days ago, to get home by boat without any neighbours to note arrival. No means of communication existed. My fears reached fever pitch before I caught a chink of light from the back room as I ran into the croft, and down to the house, breathless and tired, to find all well. The tilley had given bother, a candle stood jauntily in a tin lid on the table beside Betty who sat knitting and wondering if anything had happened to me to explain why I was so overwraught.

Our last visit to neighbour Dolian over in Glen Cannich turned out to be memorable. Again to collect straggler sheep, myself and the boys set out to visit this hardy shepherds hideaway, the nissen hut bothy of Benula. It lay on the shores of Loch Mullardoch, twelve miles from us by way of Bealach Toll a' Lochan, the high pass between Sgurr na Lapaich and Riabhachan. We arrived over there late one afternoon to find the shepherd infusing the umpteenth tea of the day. Hard by the hut to our surprise for the season wore on, was pitched a hiker's tent. Packed into the bothy hosted by Dolian and being regaled with tea and a mixture of elaborate stories were four English hikers. This ex-army hut could not be described as salubrious even amidst the primitiveness of West Benula, nor had any effort been made to redecorate or at least erase the most lurid graffiti which soldiers' sex starved imaginations could devise. A perusal of this art form itself took some little time for any who might be so inclined. No matter, Dolian bade us welcome in his grandly magnanimous style and tea in cracked mugs was liberally dispensed. The southern hikers sat ranged open mouthed on a bench at once enraptured and engulfed by a brogue filled flow of west coast humour. In an accent specially tuned for such occasions the 'singing shepherd' as he styled himself, crooned stories which ranged from the corridors of Buckingham Palace to the back of an unseemly pub in Fort William.

The conversation turned to music. In a trice Dolian, from behind a cushion, dragged out an old set of pipes. The narrow tin bothy soon reverberated to March, Strathspey and tramping feet. The MacKays, not to be outdone, took a turn. The English hikers after some ten minutes of this treatment looked visibly shaken and between a change in performers broke in to thank Dolian for all his kindness but indicating they were going to bed. With an imperious nod he blasted them out to Scotland the Brave. The making of supper, fried venison chops, did not halt the performance, for an amateur piper with the reed once between his teeth is hard to stop. Sometime after one o'clock the musical trio, tired of their recital or perhaps out of tunes, retired to various corners. Breakfast was to be early, for some reason which I don't recall.

Sure enough Dolian came awake before six. However instead of reaching for the frying pan he, presumably still gripped by some musical infatuation lingering from the recent performance, caught up his pipes. We rose at his opening blast. Soon the smell of the frying pan full of ham and eggs mingled with *The 79th's Farewell to Gibralter*. The entertainment continued during breakfast and only with reluctance did we leave the bothy about eight o'clock stirred by that splendid tune *'Leaving Glenurquhart'*. Having forgotten the ensnared English hikers up to that point, our memories were jogged as from an unzipping slit in their tent appeared a pale and haggard face. It blinked. Dolian, at once the cheerful host spoke down "Well boys, and did you's have a good night?" The face withdrew without comment. "Ach well" Dolian turned to us "didn't I tell them the ground was pretty hard".

The "white shepherd" became our friend in October. This old shepherding term referred to the line of snow which would settle for the winter at about 1,500 to 2,000 feet. Some mornings after a night of sleet we looked out and across to the bold form of Sgurr na Lapaich to see the snow down at its feet. The day would clear with a little sun and the "white shepherd" creep back to his heights his job done, the straggler sheep being forced down from many an inaccessible rock and lofty corrie, appeared without excuse for having dodged all the summer gatherings. Some proudly had a fine lamb running at heel, long tailed and lacking any stock mark. Worst of all, if a male lamb, he might be uncastrated and a strong tup lamb left out on the hill unknown to the shepherd could start mating too early in the season. Most ewes were ready for the tup in October, and a lamb or stray ram running about could result in the lambing season the following spring being in danger of beginning far too early. The consequent loss might be considerable, and this important husbandry feature required special vigilance. Other ewes hunted down by our friend 'the white shepherd' often carried their wool clip and big rolly balls they looked with the new growth pushing up under the old coat. Occasionally sheep came in with two or three season's fleeces, one on top of the other. The mystery of where these elusive sheep lived, out of sight summer and winter, could never be solved.

By the end of October the breeding flocks had been dipped, worm drenched and clearly keeled with fresh bright marking fluid. The smell of dip, if they had been through the dipper on a dry day, remained strong on their coats for the rest of the winter. Dipping days held a special terror for the ewes. They played every trick to avoid being put into the pen beside the dipper. Their final desperate act was dumb obstinacy. Nothing for it but brute force and a heavy drag for the shepherd. Care in not breaking off their horns during such heavy handling was essential. If the horn was being used to manhandle a recalcitrant beast then the ear gripped in with the horn helped to take the weight. In spite of precautions, horns broke off, especially in the cross bred sheep, whose growth was not as strong. A gruesome sight, I hated to see it for you looked down a hole right into the suffering beast's head. I smartly filled up the gaping orifice with tar.

At the dipper it was a lift and a heave into an opening which projected the panic stricken animal into freezing water for a full half minute. The brave individuals, who had been through the process on too many previous occasions, took an almighty leap, in an effort to land at the far end of the bath and thus shorten the torture period. The resulting colossal splash rightly infuriated poor Betty. Stationed at the side of the dipper with a forked stick to push each sheep's head under water she received a face full of the vile dirty, stinging dipper wash. The work was best carried out wearing coat and leggings. Steam from trembling sheep and sweating men rose over the sheep packed pens and out into the thin sunshine of chill October days which often cleared away to frost by the evening. An early start gave the soaking ewes a chance to dry before such a night.

Driven back to the hill after the October dipping, the flocks were left quietly grazing the lower grounds to help them put on a little condition and give heart before the onset of winter. Our rams spent the summer down on the strong lush pastures of the home estate, but about the middle of November a lorry load of these fat well holidayed gentlemen arrived at Monar. On rare occasions we spared the tups a walk up the lochside by loading them into one of the boats and sailing them up to their work. However as this practice made so much mess in the boats we generally opted for the long walk. All the tups would leave Monar pier with great vigour and enthusiasm, but shortly any with excess fat began to show signs of distress. Panting and tongue out the very heavy boys fell to the back whilst the thin, spry individuals, scenting the ewes, galloped on ahead. It became a difficult drive. Forty tups, half keen as mustard to get ahead, the other half wishing they could go home. Sometimes a really determined ram would just break off and go his own way. Neither dog nor man had any chance to stop him. To make the job easier in latter years I took down a cut of ewes to Monar in the boat, mixed them in with eager tups and then set off up the winding lochside path. The contingent got home in half the time.

To help with the impregnation of the gimmers, (young maiden sheep going to the tup for the first time), we made a practice of clipping the wool from their tails. The tups would also have their bellies sheared clear of wool. This latter exercise was as much to facilitate the ram's movements under snow conditions as to help him in serving the ewes.

In 1956 I put the tups out on the 15th November in conditions of singular severity. It started to snow quietly during the afternoon of the previous day. The portents had been ominous, a strange stillness enveloped the glen. Our world seemed to hold its breath. First morning rays had fired the sky an ominous orange-yellow and the daylight began to fail by midday. The tilley which had to be lit early, shone from the window as I gave the hungry tups a small feed of hay and bruised oats out on the croft under the straggling birch trees. I looked west as the flakes, small at first, puffed down the glen, the sky now of uniform steel grey shaded the hills to charcoal black. Not a cloud as such was visible. In one vast blanket the heavens looked heavy and solid. For a moment the air on my cheek seemed to warm as a tiny breath of wind hurried on before the first few straying flakes. A storm of unleashed power with all the stealth of a stalking giant seemed to be creeping in on us. On the flats not a hoof was to be seen, the deer must have climbed out to sheltered haunts earlier in the day. A lone hoodie crow silently skudded down before the now freshening wind, doubtless making for the pines of Reidh Cruaidh.

I busied myself about the byre getting in the cattle early, putting hay into the hakes up above their heads, dry rashes and bracken under their feet for bedding. The first rattle of the corrugated iron sheets at the corner of the byre told of wind to come and the blast was not long delayed. I noted the barometer had fallen near to the 28° mark as I went in for an afternoon mouthful of tea. The tups having finished their bite of hay stood in a tight group behind a low line of dyke which ran out from the kitchen garden. I went out for a last look round. The glen had vanished before an advancing white wall. The wind, now blowing hard, snatched at the last of the yellowed birch leaves, caught the white strands of dying grass on the bare croft, threshing them against the dyke side. At last, almost in relief, came the snow, blasting, stinging with an eye-screwing ferocity. The world shrank to only a few paces wide and was suddenly dense with choking particles which filled the mouth and cut the breath. In an instant cheeks and ears lost feeling, the hair on my head became coated and caked, heavy. A turn of the head snatched the breath away. A savage blizzard, the most dangerous of all weather conditions, held us in its merciless grip. I made for the house. Betty was making up the fire, stacked peats lay to one side of the range, cut wood to the other. Welcome flames danced and flirted as the blast pulled them up the chimney. I scraped quickly melting snow out of my hair into the sink and had a rub down. We settled into one of those many long winters nights, the gable, standing squarely west facing the full strength of the storm, trembled at the battle that raged. Our tiny window frames captured and built snow bridges across the wood laced panes. Supper over, I lay feet up on the long couch to avoid the draughts, reading enjoyably as I listened to the fury outside.

It was late morning the following day before the storm passed as quickly as it had arrived. We stepped out into the type of world that would be faced for the three or four months of each winter season. It was like leaving a cave mouth as the snow fell from the door with its first sticky opening. There lay a world untouched. Glistening in moulded curves and shapes with a purity of forms and profile about house and shed. An art form, subject alone to the whim and discretion of the elements. How paradoxical that forces at once so dangerous and destructive could fashion such architectural elegance of beauty and lightness. All the world's ugly edges were rounded, smoothed and filled. I looked down to the Lodge, oddly shrunk amidst a circle of deep green pine branches held high on invisible trunks. Its windows looked out from half covered walls like hollow dark eyes. Away to the horizon every hill lay reshaped, wrinkled features of rock and watercourse recast in a new great mould. Below, the surface of the loch, now still, took on the greyness of yesterday's sky, cruelly cold with more than a hint of ice to come.

Like us, the wild life only now peeped out. A raucous croak told me that the ever hungry hoodie crow was up and about. Already bird and mouse tracks cut dainty steps in seemingly aimless directions. The children couldn't get dressed and out fast enough. Gloves and duffle coats, scarves and wellie boots, as excited as puppies, they stumbled and fell in my footsteps all the way up to the byre, revelling in a new playground with so many possibilities. Deep drifts lay behind each sheltered spot, the ill fitting byre door had allowed driven spills of snow to funnel in. An unsightly mixture of brown dung and blizzard lay behind the two cows. They rose hind end first and looked round as I cleaned each stall. A forkful of hay into the rack and the steaming milk soon filled the pail. The smell of the byre seemed so much stronger. I realised the outside was without scent. Only the hens looked miserable, still on their perches as the children threw them a scoop of corn from the bin in the barn.

The children wanted a snowman, so taking the byre shovel up to the house I fashioned, with difficulty for the snow was fluffy fresh, a huge snowman. Hats and clothes came out to turn him into a smart guardsman.

I went in for coffee leaving the laughing, racing children playing outside in their new found pleasure land. Hector came running in after a little to tell me excitedly "Come out and see the wee birds Daddy". I stepped round the back of the house to find a large flock of snow buntings flitting and swooping about the croft. Almost white themselves in winter plumage, they were not unlike snow flakes as with light flutterings they picked up any blown grass seeds which lay on the white mantle. Later during December the children were taken into Inverness on a day's shopping expedition. A large store happened to have an aviary of budgerigars as a Christmas feature. Wee bright-eyed Hector spotted them and came running across the shop "Mama, Mama come quickly and see the snow buntings" he called to Betty much to the assistant's puzzled amusement.

The tups now made tracks about the croft, the snow capes shaken off their backs. They were a mixture of both Cheviot and Blackface breeds, and whatever the conditions it was time to have them out with the ewes. I gathered the group together and forcing the west gate open through a drift of light dry snow, dog, shepherd and tups set on the task of forcing a way up the glen. The drifting helped to some extent for it allowed the tups to pick a route following the stretches blown bare of snow. The glen looked so empty, only the stone fank and trees about the Toll a' Chaorachain burn stood out in stark contrast. A strange hushed world of hollow silence. Not a sheep nor a deer in sight. The tups now tiring were becoming increasingly difficult to push further forward. I sat to spy. There to my relief I picked out a large bulk of ewes gathered together below the cairn on Sgur na Conbhaire.

Dividing the tups at the old stone fank made Nancy work hard bouncing through drifts, for as half the fank was choked to wall level there was no chance to use the pens. After much struggling I separated out the Cheviot tups and hounded them stick and voice, up to where the cheviot ewes now spreading out a little, were scraping with their front feet down into the snow. Once the tups smelt the ewes they were off and I turned back to deal with the remaining Blackfaces.

The south side of the Strath fell into shadow and only after much searching I saw that the bulk of the cross bred ewes had gathered together on the exposed ridge below Corrie Dun Mhor. Only the grimmest determination on my part got the Blackface tups to the river and across water that day. Wise animals, they knew instinctively that a wet coat would soon gather up balls of snow on the ends of their long sweeping fleeces. Finally, down beside the river at a reasonable crossing point, I caught one of them on the edge of a shingle spit which ran into the sullen water. No option remained, I took him by his strong curled horns and dragged him into the current. Before getting over my knees in the freezing water I let him go. Straight back to his friends he scrambled. Nancy held them with difficulty down on the gravel. I dived and grabbed another one. Not hesitating this time I waded in up to the waist. The water swirling around me felt decidedly cold. I bawled at the dog to push them on "stick in Nancy! Had on, had on, good dog." In desperation I encouraged her to attack the tups facing her with bold eyed defiance. The swimming tup I held until he saw my intention. The current took him and with a few snorting lunges he made for the opposite bank, to stagger out under the weight of his wet coat. Now to deal with the remainder. I pulled in another one. Fortunately the sense of my determination seemed to dawn. In a reluctant group they surged into the river, swam as the current took them and struggled out on the far bank. After much coat shaking and thoroughly disgruntled they began to make up towards the ewes high above them. It remained bright though the sun had long since set, my breath hung like a cloud in a fast freezing air. Numb beyond feeling, I emptied the water out of my rubber ankle boots and standing in the snow to wring out my socks I sympathised with the tups. Catching up glass and stick I ran smartly home. Perhaps it was the feeling of returning warmth in tingling limbs, but the sheer exhilaration of wellbeing, of vigorous health in a clean, wholesome environment came stronger than I'd ever known on that crisping evening. Like an excited wild animal I leapt through the snow drifts down the path to a meal I would never be more ready to enjoy. Warm again, the tups cold swim crossed my mind "A good job you kept your feet in the river, Thomson boy". Inwardly laughing I turned to speak to the dog who faithfully loped along behind. Catching the affection in my voice, she gave a wag of her bonnie bushy tail and smiled back through her bright eyes.

Offsetting the rapid thaw and dramatic flood, particularly nasty habits of November's weather, keen frosts had not yet set in for the winter at this stage. Their steadying effect on thaw run-off was missed if a sharp climb in temperature, invariably accompanied by heavy rain and south westerly wind, occurred after heavy snow fall. Two or three feet or snow might vanish overnight causing much damage to ground surface, piers and fences. Of all weather conditions perhaps a thaw gave us the most unpleasant conditions; our world looked as miserable and bedraggled as a moulting hen. However once December came in, the weather took a sharper turn, the air became less moisture laden and more brittle.

In keeping with the ancients we marked the point of the rising December sun. From Strathmore its path skimmed the tops of the Kintail hills to vanish during the shortest days, at two in the afternoon, behind the great blocking bulk of Meall Mhor. During the fifties we had late November and early December snowfalls followed by clearing weather and high pressure, which gave rise to some exceptional frosts. No sooner had the feeble sun sunk into Meall Mhor at two o'clock in a brief afternoon of long shadows and snow brightness than the air would snap. Nostrils began to stick as one breathed in; bare hands, although not cold, would become glued to any metal object which they happened to touch. Walking over snow fields became easy as the snow quickly carried one's weight. Old footsteps turned into ankle twisters and were better avoided. The whole atmosphere became dry, brittle and extremely bracing. Nor, provided you were moving, did it seem cold, but to get wet crossing a burn or down at the boats swiftly resulted in iron clothes.

The pine trees at the Lodge standing out a welcome green, their feet deep in snow, emitted a soft crackling as the deep frost took its grip. Any branch carrying too much snow would snap suddenly with a metallic crack. Walking through the wood to the pier, as I did most days, it always amused me to notice a cone drop here and there for no apparent reason. Looking up for a cause, the crossbills could be observed working away at seed extraction from the high cones. It is a finch–like bird with the distinction of having sizeable mandibles which cross over one another like a pair of dentist's tooth extraction pliers. They were not common with us but the rattle of falling cones gave away their pine top industry.

Provided we had not total snow cover, this dry alpine weather suited sheep and deer. They could be seen working up and down the bare snow free ridges and making themselves paths from one feeding spot to another. It was not considered policy to commence any hand feeding of sheep too early in a season as they became lazy and hung about waiting for the next meal, actually losing condition unless supplied at totally uneconomic feed levels.

Prior to one great frost spell which followed a heavy snow fall I journeyed over to Pait for an evening. After pleasant hours of chat and music I set out to return home. No moonlight was needed, the snow gave off reflected starlight. Slowly and dimly at first the northern sky, as I crunched down to Pait boathouse, began to weave with searchlights. Momentarily I paused to wonder before recognising the Aurora Borealis. Strengthening, the colours shot in vivid lines. Red, green and yellow pinned to a deep blue black sky. More entrancing the weird light fell on the snow, which turned green at my feet. Until I felt chilly nothing could pull me from such an extraordinary display. The familiar hills mimicked Christmas cakes with multicoloured icing. Such phenomena and circumstances are rare and I made for home with a sense of wonder and privilege.

At the first few days of deepening frost the burns would narrow until all water moved under ice. I was then forced to cut a hole each day with a spade to allow the cattle to drink. The thirsty beasts only got out for an hour at midday and quickly learnt to take in enough water during two or three drinking bouts to suffice for the remaining twenty three hours snug in their poky quarters. I sometimes wondered how they occupied their minds – presumably thinking about the next feed for they were always pleased when I rattled in with a scoop of oats, udder cloth and milk pail each morning. All general jobs, wood cutting, tilley lamp filling, cattle and sheep feeding or anything else had to be compressed into four hours. Conditions dictated it be so.

The head of Loch Monar would quickly freeze, perhaps after three nights it might be strong enough to carry a human. Much deeper, the main loch froze differently. Ice which grew out from the loch sides and bays quickly became solid, but the main sheet stretched out day by day as the weather held hard. Boats became solidly icebound. Mail collection involving a fourteen mile hike by foot and carrying any parcels home we reduced to one day a week.

Once the ice was holding, out came the ice skates and sledge. I had never done more than roller skating before but eagerly screwed a pair of ice skating irons onto an old pair of boots and found the action required not dissimilar. The MacKay boys had skated previous winters which gave them an advantage on days when both families, including Mrs. MacKay, joined in a game of ice hockey. Spectacular sledge rides were most in demand and Teenie by no means old in spirit, joined in all the fun.

Going down for the mails, we found skating twice as fast as walking. The sledge or a branch served as transport for parcels and supplies. Not taking any risks we kept reasonably close to the loch sides though we crossed all the bays and cut corners. After a fortnight of such keen weather, Loch Monar lay frozen end to end and we knew that except for an exceptionally mild spell, it would remain closed until the spring. Slowly the level of the loch began to fall underneath this great cast iron skin. I was to learn more about the massive power of ice when boldly crossing the loch to the MacKays one night, cheerfully avoiding the discipline of rowing or perhaps a wet pony ride. I took a route not too far away from the sandbank rather than walk pier to pier. I had never experienced the brilliance of the stars to such a degree, noting that without human-made lights to detract, their brightness was ample for night walking. Many changed colour as they sparkled. Nature's downtown disco effect I reasoned irreverently. Head in air, I strode freely on, taking deep breaths of intoxicating air. Crack, a sound like a lightning strike split the air. From under my feet a white mark flashed out across the oily black ice. In some haste, I made for the sandbank. I realised then that as the loch level fell so the ice must settle into lower positions. When sufficient pressure built up, it shattered, zig-zag cracks shooting in all directions. The sides of the ice sheet soon became sloped and dangerous. At intervals, our silent world resounded and echoed from these ice splitting reports.

Once January came, sheep feeding in these conditions was a daily job. Fresh snow, during a slightly milder spell, broke the deep frost one year and fell softly with almost no drifting. Barely a land mark remained. Loch Monar still solid became a white blanket from end to end. To our annoyance snow on the ice put an end to skating. But by compensation it at least gave a sound grip when fused with the top skin of ice.

The hills showed only ripples where stone and corrie were known to be.   I set off that morning, as the snow eased, to start bringing the ewes out of the glen and down to ground more convenient for supplying some hand feeding. Below the Sgur na Conbhaire cairn, at the edge of the path I stopped at a favourite shelter stone. 'Surely no ewes will be west of here', I thought, taking out my old two draw telescope and commencing a search. The object lens steamed up. I cleaned it and spied methodically over the hills to the west. Nothing moved, a white empty untouched land, the only life under the snow. Carefully searching I let the glass follow down the burn till just below me, perhaps a hundred yards ahead, to my astonishment an eagle came into the circle. I had not noticed it when I first walked up and this puzzled me for the huge bird was plain to my naked eye. Quite engrossed it tore at the tail end of a newly dead deer, a six month old calf I judged from the size of the carcase. My first impression was of the eagle's great size as it used full wing span to help it's work. The bird stood gripping the haunch with widely spaced talons and with a vicious tug wrenched away at the meat with side levering movements of its head. I watched for at least five minutes, impressed by the primitive ferocity of the scene.

The regal bird, cruel yet dignified, looked up from time to time. A massive imperiously hooked beak held with haughty arrogance, the eagle glared about the scenery with the confidence of its contempt for lesser life.Light coloured about the chest, the serrated back feathers of dark brown, its beautiful plumage was in splendid contrast against the unvarying white background. Bright red blood spattered freely around the small miserable carcase from the hole into which the rapacious bird greedily dug. I wondered if in fact it had killed the calf for the death lay below Corrie nan Rannoch where during the stalking season I had watched a pair of eagles driving a herd of deer. On that occasion they were unsuccessful but certainly, as I had watched them swoop violently down upon a small herd of grazing deer it was the birds' clear intention to frighten a young beast over the cliffs. Whatever the explanation of this death it was getting cold and I rose to leave. After a moment the eagle saw me. It flapped off the carcase in dismay. Quite ignominiously the great bird appeared unable to fly. It commenced to flap, waddle and jump as fast as possible up the nearby hillside. I couldn't help a jeering laugh. The King of Birds had overdined and couldn't rise from the trencher. Had I been equipped with a net its freedom might have been in jeopardy. Before enough elevation was reached for a launch it lumbered upwards at least an undignified fifty yards then made a clumsy, wing threshing take off.

I turned my attention to moving ewes down the glen and after walking several hundred yards down the path I stopped, thinking to spy again. I didn't use a case for this telescope, merely a sling, and to my intense annoyance the object lens of the old glass was missing. I could not have screwed it back correctly after the last cleaning. It was bound to happen sometime, but even though the threads were worn out this seemed little consolation. Twelve inches of fresh snow lay on top of the previous fall, so my footmarks were plain as I looked back. For some reason, although I said to myself "don't be stupid, find a two inch lens in a snowfield", I did retrace my steps. I hadn't gone more than half way back to the spying stone when instinctively, most uncannily I paused. One more step and I put my hand into the snow. No searching, straight down, out I took, although completely buried and concealed, the missing lens in its brass ring. Water divining I know is not an uncommon faculty. The feeling, when working with a bent wire that one is about to come to water I have often experienced when searching for field drains, yet I have never been able to explain finding my telescope lens that day.

Getting the sheep moved to lower ground took several days. Due to the depth of snow, forcing the sheep along became impossible. They had to move in narrow tracks which followed the best route. Eventually, after some hard days the Cheviots arrived down about Strathmore. Iain came over to help me on the final mile to the hay shed in the wood at Reidh Cruaidh. This stretch, though drifted level, hid a mass of peat banks and gullies. The job looked difficult, both for shepherd and sheep. I left the ewes handy, they spent the night under the trees at the lodge. Iain walked over the loch the next day bringing Dandy with him. The frozen loch now lay below a snow cover and the pony seemed completely unafraid. Knowing Dandy's unfailing wisdom and common sense in all matters relating to where and where not to put his feet we were suitably emboldened when down to the boathouse we drove the sheep and out onto the ice. Iain had put a little hay down in a trail out onto the frozen waste. The wind will puff or eddy in settled weather, and it now started to blow in a tantalisingly mischievious manner. I had noted such tricks as a feature of its behaviour on other occasions. All the sheep, hitherto strung out following the hay trail took fright and instinctively rushed together. We held our breath. "Pick a fat one in case we go down with them" Iain shouted to me in a half serious voice. It is unwise to concentrate the weight of three hundred and fifty ewes on one spot of ice. Into my mind flashed a horrible vision of a huge slab of ice upending and sliding men, dogs and sheep into the water, closing neatly behind us. Indeed the ice did creak and grunt but after hurried consultation we agreed it would behave that way even without a twenty ton burden of sheep. No option but to take the risk, we forced the sheep along. "Sir John won't be pleased if that lot goes down" I shouted back to Iain. Apprehension still remained until the flock arrived safely at the wood. In the fashion of sheep they scurried about hungrily devouring the bales which we cheerfully spread before them. Ignorance had its bliss.

Moving the Strathmore ewes on the ice, though dangerous, struck us as particularly easy compared with the struggle in taking down the Pait ewes to safety after one exceptional spring storm. Late in March came a storm of heavy snow out of the south west. Falling on frozen ground every flake counted, for none melted. The day turned dark with a black sky to the west, the flakes that fell were of surprising size, "like falling raffle tickets", I joked to Betty as I came in from the byre. In the space of a few hours a fall of many inches accumulated on the hill. I feared it would soon be a dangerous depth as every few hundred feet of height made inches of difference. My ewes were down on the loch side, so I felt unconcerned knowing common sense would draw them to the wood and shelter. Different for the Pait Blackfaces. To eke out feeding the boys had let their sheep go as the weather eased early in March to the nearby hills of Ben Bheag and Cruachan. Never, never trust the weather. It was as though the elements were able to read man's actions and act in the most contrary fashion, The snow fell deeply. Mercifully there happened to be little wind behind the storm or the loss might have been great.

Next morning cleared to reveal an eighteen inch fall on the croft. I knew well it could be double out on the hill. The clouds blew clear from Riabhachan as I rowed across to Pait. The sun broke through, climbing high on its spring shedding arch, a lighting effect that turned the whole scene a brilliant glaring white. Every object covered, there was not a change of colour to rest the eyes, as we were soon to painfully discover. I made my way slowly up the road, knee deep in fresh soft snow. Up at the yard the MacKays had out their dogs which romped about like schoolboys seeing snow for the first time. "They'll be more sober by the end of the day" I observed to Kenny eyeing the prancing dogs. How right that turned out to be.

Some of the ewes could have been smothered due to the thickness of the fall. The boys showed concern as we slowly picked our way towards the hill, falling at times, for it was difficult to see due to the intense glare. Not half a mile out, we cut across a mass of churned up snow easily thirty yards wide. Only red deer could tramp such a swathe. We realised a massive movement of animals had taken place after the snow stopped. The great exodus led away westwards. "The deer are making out for lower ground in Kintail" reckoned young Kenny "that can't be a good sign". Indeed the deer read the weather correctly and Kenny's deduction proved equally accurate, for the snow lasted, with little break, until the middle of April. We pushed on, climbing round the shoulder of Ben Bheag, floundering unexpectedly into deep soft drifts, searching with eye and glass. The dry snow did not cling or make one cold, whilst the warmth from the sun and its harsh reflection on the snow, made our faces begin to burn. Turning, it amused us to look back on our tracks, they meandered about according to where we thought the easy route up and round the hill might lie. Here and there our misjudgements showed in great flounderings, and it was fortunate we hadn't fallen into some deep but concealed water courses. How oddly the tracks intruded upon a truly snow clad world. We rounded the sloping shoulder and much to the boys' relief, as they had rightly judged, we came upon the bulk of the ewes huddled together in a south facing corrie. Easily seen blue keel marks and black coloured faces stood bright against the unvaried background.

By now the sun blazed down out of a sky without a trace of cloud. Midday we decided and time for a piece. We stood eating not far from the sheep who realising full well the difficulty they would have in moving, seemed to have lost fear of us and the dogs. Our crusts went to the hungry collies and we turned to face the task of getting down the beleaguered sheep. The first move showed us how bad conditions really were going to be for as we circled behind the group odd outer ewes started to jump aside in an attempt to escape us. Immediately they became stuck and stranded. Fleeces spread out to their sides, they attempted a few ineffectual struggles. At once we stopped trying any force on the flock. "We'll have the whole lot stuck if we try and move them off their tramped area" shouted Iain. We held back, quietened the dogs and took stock of the situation.

"The only way might be to tramp out a path to the lower ground" suggested Kenny. We pulled out some of the ewes stuck by our first efforts. Laboriously, for fifty yards we tramped up and down in an effort to make a passage-way across to a long ridge with less snow depth which led down towards the flatter ground. A narrow snow walled passage, twisted over to the ridge after ten minutes tramping. The sheep looked on stoically as we reconsidered our method of attempting to move 250 stubborn blackfaces. This could not be done en masse. We cut off a small group. Stupidly they jumped out of our strenuously prepared path into the deep side snows. "Pull one or two along and hold them in the track" I saw what was happening and called advice. The boys grabbed a couple of ewes whilst I forced a small group in behind them. We were now hot and sweaty from our efforts, in spite of the hard frost, which rapidly set in at this height as the sun dropped behind Ben Cruachan.

The miserable dogs were on the point of giving up. Bold Shep, becoming indifferent to the problem had stopped barking. Nancy, normally willing to her last ounce of energy, tried to avoid my eye and hid behind sheep which ignored her closeness as they had never done before. Another ineffectual dive into the flock resulted in more becoming stuck. The point of total frustration for man and beast seemed close when at last the ewes themselves decided to follow my lead and began to string out in single file. Pushing head to tail they filed down the winding escalator. After one hour of heavy work we had the flock strung out across three hundred yards of track as stumbling up and down each section we goaded the bemused groups of sheep with shouts and blows. Any we gripped to hold or lift would leave a tuft of wool in the hand. Heavy in lamb at this time they were losing condition and fleeces suffered. The dogs finally gave up. We had progressed in this fashion only quarter of a mile. The purple sky to the east indicated a harsh frost and it began to bite into our hands. Faces felt burnt and the coldness after sweating sent a shiver through tired limbs. "I doubt there's no more can be done tonight boys and I've still my ewes at Reidh Cruaidh to feed". The boys agreed. All of us and the sheep needed a rest. At the end of an exhausting day the flock remained strung out in single file on the freezing hillside high above Pait, making a track as we headed down to the hill park path.

Over our shoulders the great spread of the Riabhachan face lay smooth and remote. Its contrast with summer days seemed hard to believe. Slowly, with imperceptible stealth the final rays of a garish sunset turned the rolling snow wreaths of this sleeping giant from the faintest rose petal tint a deep salmon pink. A last look back to the line of hungry sheep high on Ben Bheag told us they were in for a hard cold night. Not waiting at Pait, I rowed hurriedly down to the loch side wood where my cheviots waited to be fed under the dark sheltering pines. Frost sparkled everywhere and pulling back up the loch for home, new ice crackled under the bow whilst the oars left a pattern of holes in a quickly forming ice sheet. Truly tired that night, I walked slowly up to Strathmore, the tilley shone, the snow began to crunch and my eyes felt as though they were full of sand. I guessed I had a degree of snow blindness. Getting the Pait sheep down to the hill park took another two days of exhausting work, but fortunately the weather clouded a little and my eyes recovered.

Only on rare occasions did the deer appear to migrate in the way they had done during that heavy March snow. Like ourselves they saw the winter through, sometimes no doubt longing for the spring. The severe conditions we experienced during the late fifties and early sixties took a toll of the deer stocks. Any calves which failed to be well reared over the summer did not survive. It was largely due to high calf loss that herd numbers fell. Working daily about the hill it was interesting to note how these severe conditions were reflected in a degree of tameness. Coming onto hinds and calves in a poor nutritional state we found them reluctant to move. The poor starved calves, critically thin, their coats rough and staring walked or rather shambled after their mothers with necks stretched out and unbelievably thin legs. Likewise the stags by late December, following the rut, showed signs of poorness, but their death rate did not reach its peak until the spring grass appeared. If the growth in May came suddenly then large numbers of stags succumbed, presumably some form of grass tetany brought on by a combination of their greed and a rapid rise in nutritional value of the feed they consumed. On occasions the odd stag, using a hard snow bank, would attempt to jump over the deer fence into the croft. Unless he made a clean leap his back legs would slip between the wires which twisted round as he fell. The unfortunate, hungry brute would hang head held down as securely as though in a trap. Releasing a living stag from this position was an impossible task and I was forced to destroy them. I cut off the entangled legs and pulled the carcase down to the dogs.

Generally our dogs fed on the left overs from our own venison usage but if a stag happened to be shot for such a reason then with canine greed the carcase would be ravaged. For days the dogs would be busy, tearing at it, eating huge bellyfuls and burying special morsels. It gave us great amusement, to see Shep, tail up toddling off, as he thought unobserved, carrying a deer's leg to some secret burial spot. I would shout to him from my hiding place perhaps behind the byre door. He'd pause momentarily, look round, furious at detection, then change route and carry on, muttering to himself. "Confound it. Damned cheek interfering with a chap on an important mission". Nancy, by far the cleanest dog we ever had, was not above sneaking off to hide a delicacy, but would return nose and face all black and dirty completely giving the show away. Her look of guilt added to the comedy. "You filthy dog, get away and wash your face" I'd scold with real severity. Head down she snuffled away until my laughter brought her running back tail up and happy. I sometimes wondered who teased whom.

In spite of great weakness amongst many of the stags at this low ebb in their cycle, the life spark was not easily extinguished. I surprised a stag one February afternoon when walking home from Monar along the loch side. He had been lying behind a knoll perhaps twenty yards above the waters edge. A smart south west wind in my face, I followed the path around a curve to step within yards of this poor emaciated creature. I recognised from the ragged, tufty nature of his dull coat that his body condition must be miserably poor. Stumbling rather than springing to his feet at my close encounter he ran the little distance to the water rather than tackle the climb back to the hill. The shock of my dramatic appearance was so great he floundered into the loch and commenced to swim out into the dark icy February waters. Head held high but back barely above the surface he swam with more speed than I expected even causing a little ripple to follow behind his progress.

The nature and performance of weak sheep which take the step of seeking protection in the watery element is standard. They swim from the shore perhaps forty or fifty yards before becoming disorientated. At this point a suicidal fatalism typical of sheep takes control. They begin to swim in ever decreasing circles. Unfortunately the end never varied. After fifteen minutes down they went and there was little the shepherd could do unless a boat lay handy.

The stag as I watched, struck out, emaciated though he was and swam a straight course, crossing the loch at about its widest point, something over a mile. By the time he reached the centre he became a slowly moving dot and I sat to watch. Without slackening swimming speed, after twenty minutes he made the far side. The gloomy waters did not claim a victim as I fully expected. Through the glass I saw him stagger ashore for a few yards and lie down without even bothering to shake his coat.

THE CHRISTMAS PIPE

During the winter of 1959-60, we were alone in Strathmore. The MacKays now retired from Pait, lived down country, enjoying a well earned retirement in a croft convenient to village and social amenity. We greatly missed the light across the loch, the coming and going and close friendship which mutual help created.

That autumn old Kenny suffered a nasty cut on his eye brow one day whilst cutting winter logs in the old peat shed behind their house. It happened that the MacKay boys had gone to a late season's Highland games. Betty and the children, taking the opportunity of a lift down the glen, journeyed away visiting friends for a couple of days. The accident occurred in the darkening and apart from bathing the nasty long gash, the old couple took no action that night. The following morning I went about my chores on the croft in the normal way never failing of course to look across to Pait from time to time. The agreed signal between the houses, meaning 'come over at once', was a white flag hung out from an upstairs window.

About mid morning I glanced over and spotted the flag. Knowing something was wrong I quickly rowed across and walked in to the house to find Mrs MacKay bathing this very ugly gaping cut. Old Kenny, who had suffered a bayonet wound during the first world war, made light of the injury but it was clear the eyebrow could not be left hanging open. Without more ado, Kenny and I set off for Inverness leaving Mrs MacKay completely alone for the first time during her 40 years of life in Pait. The *Spray* lay down at Monar. We had only a rowing boat powered by a small outboard engine. Kenny sat on the middle thwart, back to the wind, for a slow cold run. All went well, though the day was not improving but blowing up from the west, and ugly clouds piled low over the hills. Allan the Monar keeper, on seeing Kenny, at once ran us to the hospital in Inverness where we saw him safely under care.

That night Allan and I came back late to Monar. A good gale blew heavy clouds through a very dark night. Although pressed to stay by the Flemings I decided to make for Pait and then home. A rough run for a small boat. The white wave tops curled into evil grins barely hiding their contempt as with sinister intent, behind the night's dark face, they tossed the rowing boat about with capricious ease. After dangerously reaching broadside across the east end of the loch I headed over to the northern shore, by far the safer should anything go wrong. Sure enough the outboard spluttered to a stop just off the half way house. I could make out decidedly choppy water where the river mouth entered the loch. Hurriedly I grabbed the oars and rowed for the shore, annoyed and determined the wretched engine would see me home. In moments we bumped ashore in a small bay, each wave lifting the boat higher onto the stones with a distinctly unpleasant thud. I jumped out to pull her in clear of certain damage whilst I filled the fuel tank. The wind spurted the pouring fuel over seat, engine and my legs. I cursed. It took a gallon to put in half that amount. 'Damned if the gale would beat me', I vowed, when ready to go I faced the onshore gale and poled out with an oar.

The bouncing boat drifted in too fast for even one tug at the oars. Thoroughly wet I became really angry. Cursing the gale, screaming abuse at the elements I assured them in language which surprised even myself that it was all or nothing. Wading out waist deep with the boat head on to the wind, I scrabbled in over the stern, dived for the oars and attempted to pull free of the shore in order to clear the stones before starting the engine. Would the engine fire? Rage and frustration gripped me, hardly could I turn round before the boat bumped back aground. Three attempts, wading, pushing, rowing before the stupid engine finally condescended to fire.

Wet, shivering and still muttering imprecatious about the weather I arrived an hour later at Pait to tell Mrs MacKay that old Kenny had got the wound stitched. It was four in the morning, I took a welcome dram from a relieved old lady and made home to Strathmore. Perhaps it was this accident that finally stirred the old folks to leave a home which had given them a family life of great contentment. The winter after they left was particularly harsh. Monar froze again from end to end and there was much heavy snow. I stopped bothering about collecting the mails and we were cut off exactly six weeks, without connection with the outside world in any way. It did not worry us or even seem strange. Days rolled into days. Time passed with surprising speed and certainly without any pressure.

The long hard spell of our last winter started on New Year's day. The day opened with a morning of exceptional brilliance. The loch lay serenely calm as I took the *Spray* to Monar for the time honoured social call. High snow levels fringed the top of Maoile Choill'Mhias. I sailed quietly east. Reflections in the loch before my disturbing wake broke the mirror were a perfect double. The world stood on its head. I could have touched the hill tops over the side of the boat. Russet bracken joined the golden sheen of winter-dead grass to glow from hill to water and shine up to me in a day of unusually sharp light. My eyes ranged far and wide over the brilliant view. Halfway down the face of Maoile Choill'Mhias I started, catching sight of what I took for a second to be man walking up the hill. Quickly I set the tiller and took out the glass. No, I was wrong, the movement which attracted my attention, sprang onto the lens as a large golden eagle. Rather stupidly and in a most ungainly fashion it waddled up the hillside. I put the glass down for although the distance must have been at least half a mile, the bird's antics were easily watched from the boat with the naked eye. Eventually it strode out onto a sharp rise and with two cumbersome flaps launched its heavy frame into the air, narrowly missing the ground before soaring out above the loch. It circled above me once an air current lent power to the effortless nature of the great bird's flight, before striking away east to the great pine forests of Glen Strathfarrar.

Needless to say the fine day was the precursor of singularly severe conditions which did not yield their grip until the middle of March. I came home during the early hours of the following morning as the frost began to form. Balancing boldly but a trifle unsteadily on the bow of the *Spray* I left her chugging in gear as I broke a passage into Strathmore pier with a boat hook. What might have happened had I fallen over the side did not cross my mind.

Going back to the boat later that day I realised the frost meant to take an iron grip for the head of the loch was almost bearing at the edges. I ran back to call all hands on deck in a race against the frost. Using the rowing boat Betty and I set about breaking a passage down to the corran. By now the ice was too thick for the oar to hole so whilst Betty rowed a little at a time I stood in the bow smashing all round about me using a long iron bar. Twenty minutes of hard work and we reached open water. Landing the rowing boat below the corran we both hurried back to get out the *Spray* before she would become marooned as securely as the famous *Fram* of North Pole fame. The passage, broken only twenty minutes previously was already freezing again as the launch ploughed slowly out through tinkling ice debris. Betty went home with the shivering children whilst I ran the *Spray* down the loch side about quarter of a mile to the small inlet once the landing place used by the MacLennans for their croft. In memory of Eachan Mhor, we called this sheltered little cove, Hector's bay. From this anchorage I hoped to be able to operate the boat so long as the main loch remained open. Using opposite shores of the inlet I chained the boat stem and stern and left her floating safely in the centre of the bay. Walking home in the last light I knew from the dryness of the atmosphere we were in for a cracking hard frost.

So it proved. The frost tightened its grip on us for over a week, sealing loch and boat, drying burns and vegetation but giving clean healthy invigorating weather. My daily round of livestock feeding took me down to the loch side wood to spread hay for the cheviots and also involved my going over to Pait by crossing the ice at the sandbank to feed the few cattle wintering in what had been the old home of MacRae the whisky smuggler. Pleasant easy work, I walked dry foot everywhere.

It was not to last. The middle of January saw a blizzard which lasted over two days without a break resulting in a heavy snow cover. I had attempted to go up the glen to draw in sheep as the blasting north west wind whipped in the first of the snow but soon unable to see ahead, I turned back. Even the dogs were beginning to jib on that day, their faces comically miserable as they became encrusted with freezing snow which refused to be shaken off.

The weather then set a pattern of several more snow falls followed by spells of frost but without any sign of a break. January passed into February. We had long since run out of bread. Betty baked a home made loaf which seemed to be rather better than shop bread and was no doubt a good deal more nourishing. Experience told us what the seasons could do and a very ample supply of basic foods had been stored early in December. Venison, oat meal and potatoes with plenty of milk and home made butter, we really lacked for nothing by way of wholesome feeding. I recalled old Kenny telling of great famines in the Highlands, particularly in the years around 1846, about the same time as the disaster which struck Ireland. Along with oatmeal a general practice was to bleed the cattle during the winter months to provide nourishment. We were certainly not reduced to that dangerous level.

Towards March however a break became imperative. The livestock were suffering and feed ran low. Ewes listlessly poor, in some cases began to nibble their fleeces, such seemed their need for a change of diet. The deer had mostly moved out to the west coast and life forms became reduced to the hoodie crows and ravens who never went short of a meal in a bad season. Our tracks to various operations led like highways in the appropriate directions, trampled and hard. March came in with another blizzard and conditions did then reach a trying phase. The lengthening day made the animals more hungry. Frost on snow prevented sheep and deer from digging for feed. The ewes now spent the whole day lying about the feeding areas, waiting for the next meagre ration. My cattle at the Pait fank were running low on hay. I had managed to make a small amount that previous summer across on a tiny piece of arable ground which lay on the Attadale estate over the burn from Pait. Rations for them got down to a handful per day. I realised the fear which haunted the past generations.

The change came with little warning during the second week of March, oddly heralded by what seemed oven hot puffs of wind from the south. The sky clouded and a few spots of rain fell that afternoon. I couldn't believe it, wondering how such a set world would react. We were soon to find out. All night the incessant drumming on the slates told of a steady downpour. Then came the change of sound, like a murmuring chorus moving somewhere far beyond the hills but winding closer with deliberate tread. The music grew louder, faster, until muffled at first other sounds joined the rain song. A low rumbling, odd splashes and gurgles grew in volume as the silent, winter-bound world began to stir and stretch its limbs.

Next day the warm rain drummed on slippery snow without a break. Looking to the hill I sighted the first hints of black ground on the long ridges. The thaw was gaining momentum. Down at the solid burn mouth imprisoned water broke through and babbled over a foot of ice. Rain, rain, it continued all the next night sweeping in like a warm monsoon from the south. The cacophony of sound increased, the sheep suddenly left their hay beds and made for the fast appearing snow freed ridges. Below the house I could see the Strathmore river heaving into life as it passed out over the thick ice sheet. Water now lay everywhere. The house steamed and sweated, walls, floors, clothes, all felt damp and clammy. Difficult to avoid a feeling almost of regret at the loss of an ordered world.

On the third night the wind swung to the west and began to rise in strength. The pitter patter of heavy south born droplets gave way to the familiar slash and rattle of wet weather from the west. That night the thunderous explosions started. Not the rifle sharp crack of sealing frost but duller drum-like salvos of rotting ice pressured from underneath by rising water. I went out the fourth day to a fearsome scene. Torrents rushed everywhere, over snow, down paths, across the croft, tearing down burns, filling long disused watercourses, ripping out banks, rolling boulders with massive scenery-changing destructive power. Gale lashed rains gushed in reckless union with the freed waters of the thaw spilling and spreading over the landscape.

Towards midday donning boots, coat and leggings I struggled down to Hector's Bay, fearful that the breaking ice would tear away the boat. The elements tore at my clothing as I slithered down the water borne pathway. Reaching the sandbank the ice which I had not dared to cross the previous day now parted along the line of the sandspit and in undulating movements the great mass was coming alive. The raging westerly gale forced open a passage of clear grey wind-flecked water on the Monar side of the sand barrier. Huge ice floes, in an unstoppable shambles were waltzing before the wind in a sickening ricocheting fashion down the loch.

I reached the *Spray*. She lay still immured in a circle of ice which risen free and floating, allowed her to collide with the edges of the ice pack. I contemplated the situation. Not sure what to do, aware the safety of the boat was at stake, I stood and weighed the alternatives. To leave the boat in the bay with ice grinding all around her would surely see the launch thrown upon the shore. Huge ice floes weighing tons would be no respector of chains. I saw clearly this imminent danger, for as I pondered action, the first of the juggling mass from the loch began to grind up on to the shore beside me. Massive plates of ice rose up the banks, without apparent effort, pushing stones and bank before them as they reared on end to break and make way for the next slab. The noise was awesome. Grinding and scraping and crackling stone on stone, snapping ice falling on ice. I had not witnessed before or since such a spectacle of raw elemental energy. It turned bitterly cold with more than a hint of sleet. Rain down my neck, I started to shiver. Action was essential to save the boat.

The stern chain was held tight by the weight of ice. I scrambled desperately hand over hand along the taut chain trying to keep my legs safely above the dancing ice menace, and managed to get into the boat by climbing up the rudder. Emptying my wellingtons I made a hasty judgement.

The solution lay in a bold attempt to get the *Spray* across to the safety of the pier at the new boathouse on the Pait side of the loch. A wide channel now extended between the sandbank and the east flowing ice pack. Open water was visible towards the bay I determined to break out of. The good old engine started first pull. I gave it a warming run. Casting off the bow chain by undoing the shackle, the next move was then crucial. I edged into reverse gear to ease the stern chain, uncomfortably conscious of mobile ice crawling up the protecting headlands of the bay. The chain came free. I began to prod the bobbing floes in an attempt to get the boat to swing her head out of the bay. The menacing ice hammered and ground at the boat sides. Without option I shunted back and forth in the tiny area of free water. The rudder took a splitting jerk as it came against unbroken ice but mercifully the *Spray* began to turn. At last her head pointed out and I forced through the last of the jostling pack ice and into clear choppy water.

With intense relief I opened to full throttle and headed across towards the new boathouse. My main fear centred on that great mass of ice lying at the top of the loch west of the sandbank. Might it suddenly, due to the rising water, sweep down and trap the boat against the pack ice? I kept a watchful eye on this menace, whilst romping the ex-life-boat across to safety. Too late I realised sizeable ice floes were swilling down from the mouth of the roaring Riabhachan burn. The boat shuddered sickeningly as we struck one. I feared for her planking, anxiously watching for a gush of water as I watched twenty foot floes slide along the boat side. Fortunately it must have been a glancing blow. Rounding the sheltered pier I moored her well up the jetty to allow for the inevitable flood which would rise rapidly, fed by the torrent rivers. I took a moment to consider.

From the shelter of the new boathouse I looked across the flats towards the Pait pier, the fanks and the old stone house from which my cattle, spotting a movement, had now emerged to stand bawling across at me. Poor beasts unfed for two days, I had not ventured over the sandbar due to the danger of crossing rotting ice. The wide view unveiled about me was one of incredible desolation. Through gale-slanted rain the hills looked ugly. Black streaks of bare ground ran up their slopes irregular marks criss-crossing the dull off-white slush. A fast vanishing clean image decaying to shapeless squalor. I shivered, it was really cold. Wet from getting into the boat at Hector's Bay, I now felt the numbing effects of the driving rain and the chill that always accompanies the misery of a thaw.

The sky, grey and cloud-torn in its rain laden struggle across the hill tops, was fast losing daylight. Below me, the flats of long summer days at the hay evoked a sense of the primitive chaos of creation. Both the Garbh Uisge to the west side of the fields and the Riabhachan burn beside me had dammed their outlets with vast ice floe jumbles. As a result the surging water seeking another entrance to the loch had changed course. The Garbh Uisge now tore through its old channel down the middle of the croft. Should I cross these waters and home by the sandbank or take back the boat? Nobody to help me to make the choice. I spoke to the dog which I had foolishly allowed to follow me that morning. "Well Nancy we shall need to take care" she looked up, thoroughly wet and miserable. 'Perhaps as well you can't reply', I thought, as I made a quick decision.

Down to the Riabhachan burn by the path from the boathouse we made to where the route over the waters was traversed by a wire and plank swing bridge. The fast darkening water ripped under the contraption with a few inches to spare. I knew there to be a two foot sag when one's weight reached bridge centre. Ice floes swirled down the rapids, and I remember Iain MacKay once telling me of losing a good yellow sheep dog down here one winter. Crossing the river in similar conditions the dog faltered and was swept under the ice. The mouth of the torrent dimly away to my right looked securely jammed. No turning now. I caught the dog under an arm and waded to the steps of the bridge. I watched the river for a few moments in the gloom until judging it clear of approaching ice, I clambered onto the flimsy swaying structure. Hanging grimly onto the dog as best I could, I slithered my feet out along the planks. The bridge dipped into the torrent and swung out in an arch before the racing waters. Now well over my knees at the sagging centre I caught sight of a sizeable ice floe twenty yards upstream. Never have I moved so fast. As I grasped the pillars at the far side the heavy floe snagged the bridge. To my horror it hung for a second before rearing into the air, twisting over itself and sliding over the wire bridge. I was shaken by the sight.

Now committed between two madly flooding waters I became extremely concerned. The weather did not relent. Darkness closed by the minute. I ran over to the next danger. The diverted Garbh Uisge ripping down its new course. Twelve yards wide, it must be quickly crossed. Selecting a point above where the surly mass poured through a deer fence I waded in. Nancy pulled by the neck, up to my middle, I threw the terror stricken wretch with all my force. She landed half way across to be swept against the fence wires. By terrified clawing and swimming she got out. Unheeding now of cold or caution I lunged out myself. The strength of the current threw me hard against the wires. Like the poor dog, by the grace of the fence I made it. Dripping, squelching, wet to the skin I paused to empty wellington boots, flew up to the old house, felt about in the gloom and threw the unconcerned cattle two armfuls of hay.

Alarmed at my position I called the bewildered dog through the dark howling wind, and hastened down to the sandbank. Reaching home depended on getting across. Would the ice sheet on the west side start to move? When would it decide to slip smartly over the sandy spit? Running along the firm sand out into the water, I could see the remaining ice sheet yards to my left at the point where it had ruptured open that morning.

Frightened, in desperation I waded hurriedly out along the shelving bar. The navigable channel curved ahead. Swim it? I ran back to shallow water, took off wellington boots, tied them upside down under my belt. Down to the channel, now without caution, again calling the dog, I plunged in and swam the twenty five yards. My feet caught stones on the home side. Water poured from me, relief poured in. I turned, no dog. Nancy must have turned back. Oh God, I plunged back calling, calling, then her white collar showed. She was swimming. I caught her. It was freezingly cold, I shook with cold and exertion. Black dark, on a devil of a night I heard the unmistakable sound above the raining wind, of ice on the move. There had been only a few minutes to spare. The dangers of isolation stared out into the night.

I walked up the path towards the tilley's blinking welcome. Realizing Betty with the children had been alone half the day and long into the darkness, she would not have known if I had gone the same way as MacKays' yellow dog. Next day, to our surprise and pleasure, a rucksack laden figure tramped up to the house late in the afternoon. It was Iain, with mails and bread, the first person we had seen in six full weeks.

Both family life and neighbourliness were drawn especially close at Christmas. Perhaps the simplicity and uncomplicated nature of our life style added warmth to those days but I suspect the very distance from the outside world also gave much by way of compensation. Our neighbours were greatly missed the last Christmas we spent alone in Strathmore. Perhaps the isolation became magnified when just the four of us sat down to partake in what should have been a traditional turkey dinner, grateful for the bird generously provided by the estate. I write 'should have been' because an accident befell the festive bird before it had opportunity to take its due place at the centre of the laden table.

The weather had turned blustery with unpredictable westerly gales just a couple of days before Christmas. For once I left the *Spray* at home but knowing that a collection of parcels would await me down at the old shed above Monar pier I threw a blanket tied with a strap on Dandy's back and rode down the loch-side in my official postman capacity. I here disclose the irony of being paid by the G.P.O. to collect my own mails. A happy group, off we set, dogs trotting at Dandy's heel with a zest for every scent as we progressed down the lochside on a now clearing and pleasantly breezy day. A large haul of parcels awaited my official attention. I duly stuffed them into my G.P.O. issue postbag and various other less secure haversacks. A magnificent turkey hung swinging by the legs on a string from the rafters of the old shed. A dangling ticket read, 'a Happy Christmas to you all from Sir John and Lady Stirling'. I eyed the braw bird for a moment knowing transport presented a special problem. Eventually I bound it by the legs to the outside of my assorted kit and mounting the lively Dandy we set off gaily on the homeward track.

The dogs enjoying a day of exercise and fresh sniffs, meandered behind stopping as they pleased. It turned to a splendidly brisk evening; the tail of the storm lifted to give towering silver-plumed cloud effects which built high above the raw ochre coloured hills of Kintail. Dandy felt the spirit of the day which induced him, along a good stretch of the path below Creag nah-Iolaire, to shake his head and break into a smart trot. Taking his cue a little encouragement lifted the pace to a spanking gallop and we covered half a mile in rollicking cavalry style. Slowing to a walk as the path roughened I put my hand behind to check the valuables. All seemed fine as I felt for the prestigious turkey. It had gone, I turned, no dogs either.

Wheeling Dandy we cantered back down the path, leaning into

several bends at speed. Whin bushes hid the last twist of the path in that stretch, pony's mane and rider's hair streaming we rounded it in fine gallop. There, twenty yards ahead, Nancy and Shep, tails wagging wolfed into the sacrificial bird. My rage knew no bounds. With a most unchristian shout I leapt down and tore the tattered remains from their feather filled jaws. The mouth watering bird was mangled far beyond oven condition. Withering them with grudging comment I flung it back to the secretly smirking pair and turned for home. The premature revellers were some time before they caught me up. I looked down, not really annoyed any more. "Well dogs, I didn't expect you to have to carry the turkey home." They must have smiled back and wished me the compliments of the season.

Whatever the weather, Christmas afternoons and evenings in earlier years were enjoyed with the families together at Pait. For once we went about the necessary duties early in the day. Cattle, dogs and sheep all got extra rations. It was really more by way of keeping them quiet the following day than Xmas goodwill, so it cut both ways. Good clothes, steamed dry of dampness looked smart and with the children bundled into duffle coats there would be a row over to Pait about two in the afternoon. Iain, watching for our boat brought down the jeep, as a special treat for the children. We drove in state up to the front door. This annual use of the front door betokened the auspicious nature of the day; Mrs. MacKay even waited on the step. "Here you are" as she always greeted us, then "Happy Christmas". Kenny, Biddy and her husband with hand-shaking round about welcomed us in the tiny porch. From the warm enthusiasm of this welcome it could not be deduced that we saw these friends at least several days a week. It might appear to an outsider that we had not met for years.

Crowding into the porch, taking coats off, much excited chatter started our celebration. The children always showed intense fascination in a large stuffed eagle which stood with fierce, dignified aloofness on a pedestal squeezed into the porch corner. Old Kenny had caught him quite by accident in a trap out on Ben Bheag many seasons previously. They were allowed to stroke the noble bird's glossy brown feathers. Old Kenny himself, inveigled into putting on a different knicker suit, sat in his special chair within foot-reach of the stove. Putting down his pipe and waiting till the last moment, the old chap rose as we entered the kitchen. "Happy Chrismus", there remained always a hint of his childhood Gaelic in the accent.

The smell drifting from the scullery spoke of the feed in store for us all. Tilleys lit early for the day caused the children to wonder at the few decorations pinned to the wall. They waved gently in heat rising from the hissing lamps. The small room came aglow, peats blazed up with a poke from Kenny and the meal moved to the table. "Teenie" a tireless worker all her life, talked and laughed as plates came through laden with turkey and goose. No venison today, we joked, bandying again the old stories which always seemed the funnier on such grand occasions. "My jove this is great, tasty stuffing you have with the turkey Mrs. MacKay" I said, solemnly relishing a mouthful. With mock innocence she replied "Oh I just made it out of my head" and then became helpless with laughter for she always enjoyed a joke against herself. The meal more than sufficed us young and old. Crackers, paper hats and a dram to wish absent friends, along with our own merry crowd, all the best wishes of the season. The feast demolished, curtains drawn, we sat over to relax in the warmth of good company.

Old Kenny did not need too much persuasion to regale us with accounts of his adventures in the first world war but to bring him up to date a little I would venture "Tell us about your life during the last war Kenny."

"Well boy, you see it wass like this, I wass too old you see, but the authorities knowing the value of an old soldier, they put me in the Home Guard." With a blunt bent finger he pushed tobacco into his pipe. We waited without breaking the thread. He continued "I wass issued with a tree -o- tree rifle and aal the kit as well you understand. The Major down country, one of those you know" — he indicated by the subtle inflection of his voice that it would be some pompous Englishman whom Kenny with the hindsight of his spell in the trenches, regarded privately as a nincompoop, "expected me to dress up, drill and have a kit inspection effery Sunday morning. In the event of invasion I was instructed to defend Pait to the last man" he pulled at his pipe, "Ach well, there wass only myself and her anyway, but it wassn't a bad rifle you know." We laughed and he smiled himself. "I believe it wass in the November and there wass storms for a week. I didn't bother going down for the mail, we had plenty to eat anyway. Never mind," his eyes drifted into the distance as perhaps an unspoken thought passed through his reflections "I went down to Monar on the Saturday and there be jove, pushed under the old garage door lay an O.H.M.S. telegram. Oh boy I'll tell you, it read 'German parachutists imminent, prepare to defend territory'. Hoch, hoch, I hurried west wondering where jerry would land. It called the instructions secret, but ach, I showed the telegram to her" he said, nodding across to

Teenie, who by this time was resting her eyes, being not unacquainted with the yarn "and all she could say was "Kenny it's dated a week past yesterday."

Teenie opened an eye and laughed. Kenny ignored his wife with mock gravity. "I didn't want to let the nation down you understand, but I knew jerry wouldn't come a week late and on a Sunday — it wass wet anyway. But I tell you boy" he said "you being in the army would understand. I put on my uniform took the .303 and went up Corrie Each early on the Monday morning. Not a jerry in sight for all that trouble but by jove I got a topper of a hind on the way home. It wasn't a bad rifle you know," he concluded with a wave of his pipe. We would laugh and laugh at this famous story. The children though not understanding, caught the fun and laughed with us.

To round off the evening, tea and Christmas cake awaited us through in the sitting room. A fire, seldom lit, did its best to thaw the chimney and warm the room at one and the same time. Betty summoned to the piano would lead the favourite carols before the boys turned on the old 78's and we danced in the tiny room. Not too late, it came time for home, we gathered up our belongings, coats piled on a chair, childrens gloves, Presents were few but meaningfully given. The children shyly handed over tobacco to old Kenny who quietly pressed money into their reluctant hands. Gravely they gave perhaps a head scarf to Mrs. MacKay. Goodnights were brief, for the night struck cold after the warmth of house and company. Iain ran us down to the pier and as the jeep lights turned back up the road, we rowed to the home pier in the darkness. The motion of the boat put the children to sleep and landing quietly we carried them up a happy path to Strathmore.

EPILOGUE

The year 1960 became a truly significant watershed for Glen Strathfarrar. It divided this faraway unspoilt world of natural magnificence from the developments and inroads upon its beauty and seclusion which have been brought about by the past twenty odd years of progress. Such sudden physical change and altered human attitudes should be weighed from various points of view. My experience straddled this breakaway from the last days of an essentially class based Victorian life style with its wealth based prerogatives, to the current growth of State intervented bureaucratic control, fronting a supposed egalitarianism in matters of countryside access and land usage.

We were the last family to live in Strathmore, to taste its unhurried pace of life, its calls on strength and self reliance. To actively practise the essentials of good neighbourliness and enjoy compliance with a system which had existed with only a slow rate of change over many generations. The willing acceptance and respect for values far removed from the common currency of today's social patterns was an essential attitude. Without consciously conforming, the adjustment came naturally. There was no contrived behavioural change on our part. Allowing full play to innate feeling and instinct we fell into the rhythm of the season's cycle. Accepting without impatience that our daily lives must often comply with factors beyond our control, we turned to working with the elements not against them.

The tentacles of the 18th and 19th centuries industrial revolution reached into many corners of the globe and certainly felt their way into the Scottish Highlands. Intensive crofting and transhumance to the high summer sheilings, a practice which had existed for many centuries, began to decline about the early 1800's. For largely economic reasons, willingly or otherwise, the people left their native glens. To take the place of a concentrated human element in land usage, breeding stocks of South Country Cheviot sheep were introduced. Brought from the Borders, together with many southern Scots shepherds, they spread over the Highlands in a space of thirty years. The sheep throve on the residual soil fertility built up by herds of native cattle whose numbers declined proportionally as human population figures fell. Perhaps little appreciated is the bovine's ability to unselectively consume large quantities of roughage from hill areas. This is a major factor in allowing for a more effective photosynthesis to take place on land which would otherwise degenerate to courser, less useful and unpalatable plant cover. Cattle therefore played an essential role in a symbiosis which maintained a relatively high human population density, albeit at a somewhat precarious subsistence level.

For a further fifty years after the introduction of sheep stocks to large tracts of land such as Strathfarrar, the west coast populations still adopted a form of transhumance. It involved the young and able walking, in the early summer, after completing the essential peat cutting, through to the farmlands of the Eastern seaboard often reaching as far afield as Aberdeenshire. On these large fertile farms this cheap labour force worked the harvest and potato lifting before trekking home to the West Highlands in the late autumn.

By the time Red Deer stalking became a vogue for many "nouveau riche" of the industrial revolution's profitability, the indigenous population of the Highlands had been dramatically reduced. The facts and methods by which this decimation was achieved are well recorded elsewhere. No reference need be made to the "clearances" except to say the Lovat Estate did not actively clear Strathfarrar. Indeed they actually allowed some resettlement from lands of "The Chisholm" in a neighbouring glen, where the policy adopted by this Proprietor towards his tenants was less considerate.

Suffice it to say that in spite of population and husbandry changes over a period of some two hundred years, most glens remained capable of coping with the vagaries of human attitude to land usage without any rapid or major ecological reshuffle taking place. The coming of hydro electricity generation however, constituted a dramatic change to relative stability for large areas throughout the Highlands.

Glen Strathfarrar in 1960 was under the ownership of three Proprietors. The lower end, from Struy up the six miles to Deanie, by the Spencer Nairn family. Sir Robert, a keen fisherman, assiduously guarded the privacy of this stretch. The extensive and varied centre portion running west, belonged, and still does, to the Lovat Estates. It included Deanie, then stretches as far as the watershed at the foot of the massive Sgur na Lapaich before following the Inverness-Ross-shire county boundary across to the south shore of Loch Monar at Cosac point. Further west again and to the watershed with Kintail we have the wide deer forests of East and West Monar together with Pait. They were owned up to 1963 by Sir John Stirling of Fairburn, Muir of Ord. With mutual agreement between the three owners, a gate was maintained at the entrance to the glen about a mile from Struy village. Though rarely closed the glen road was unquestionably private. After the war the road fell into a state of disrepair, sufficiently bad to dissuade all but those insouciant of their car's welfare, from attempting beyond the first couple of miles. In the fifties, tourism had not gained the penetration into the Highlands which it presently enjoys, and Strathfarrar remained largely unknown. As a result the glen existed almost untouched by other than authorised admissions on Estate business. It must be borne in mind the liberal right of "no trespass" in Scotland extends only to pedestrians. The Proprietors were within the law to maintain and close a gate against vehicular traffic.

During the 1950's investigation work went ahead by Engineers of the North of Scotland Hydro Electric Board with a view to assessing the area's potential for electricity generation. Finally in 1959 the Board decided to press forward these plans for harnessing the waters of Glen Strathfarrar. Already the two neighbouring glens of Affric and Cannich had been dammed and flooded. The final phase of the Board's overall plan was to dam Loch Monar and link Strathfarrar to this existing complex. Two generating dams were added on the lower Beauly river to complete the water control and utilisation of these vast square miles of the Highlands.

The plans, duly formulated, were placed before a public enquiry held in Edinburgh during November 1959. Though there was considerable general outcry and vociferous demand to have the last remaining unspoilt glen of the Highlands left intact, the opposition to the scheme lacked the basis of a well reasoned co-ordinated case. Not surprisingly the Hydro Board presented a carefully documented brief which ostensibly gave tangible evidence, concurrent with the general public outlook, of ensuring the furtherance of wider Highland welfare in the form of cheaper electricity. An outcome favourable to the Board was never seriously in doubt.

The three Proprietors lodged formal objection to the proposed scheme. Sir Robert took by far the most determined stand to prevent the scheme's implementation. Finally however, upon the outcome of the public enquiry, their objections were withdrawn and amounts of compensation to the Estates agreed. The Lovat Estate obtained of the order of £100,000 — this sum being chiefly as compensation for the loss of salmon revenues during the expected five year construction period. As part of the negotiated agreement it should be noted that the Hydro Board undertook to build and maintain a road from Struy up to the Monar dam. This vastly improved access they required for the construction work and subsequent servicing. On completion of the project it was agreed by the Board that the road should return to the private ownership of the Estates involved.

Public outrage developed when at the end of all work in connection with the Scheme, a gate was replaced on the glen road at Struy. Marked 'Private', it was duly locked and a gate-keeper installed with instructions to check all admissions. A system of day passes operated. Whatever wider view may be taken, the Proprietors acted within their legal rights in terms of common law and in accordance with the agreement drawn up between themselves and the North of Scotland Hydro Electric Board. Whether or not a public body should have powers to draw up such terms without wider reference is another matter.

I'm glad to say we left the glen shortly after construction of the new road got underway. The despoiling of any area of beauty is at its most painful when the heavy plant tears without remorse at hillside and birch tree. A glen once busy with pastoral affairs now witnessed a form of activity reflecting man's ingenuity and ability to turn the forces of nature to his advantage. Some of the planners however were sensitive to the beauty amongst which they worked, for a good deal of attention and expense went into landscaping the various generating stations and water tracks involved. Nevertheless it seemed that the outlook of the vast majority, public and planner alike, regarded the taming of the glen as a considerable achievement. Certainly little concern attached to the effects produced by flooding to the depth of 100 feet, some six miles of alluvial valley flats at Strathmore.

The production of a small amount of electricity now largely directed to the conurbations of the south, was regarded as perfectly adequate reason for the sterilisation of such low grade and relatively unimportant land. Much temporary employment for both itinerant workers and local labour became available, and on completion a small maintainance workforce to tend the installations was imported. Preparations in the flood area of an extended Loch Monar went ahead in 1961. The outlook of the Contractors and labour force might have provided interesting insight as they set about destroying these long settled habitations.

Larch cut from Monar yielded some of the finest timber which the Contractor said he had handled. The wood on the Loch side at Reidh Cruidh was cut and floated out by forming rafts towed down the loch by the old *'Spray'*. Monar and Strathmore lodges were burnt and blown up. Our cottage and byre, a place of constant habitation for centuries, received similar treatment. The Lodge at Pait remained just above flood level and was spared, but the cluster of houses including the MacKay's were left with water swilling about them. The spectacular horseshoe shaped Monar Dam constructed above the powerful Monar falls in the narrow gorge at the loch's outflow did in fact, when filled in 1962, lengthen the loch by some six miles. Its rate of fill astonished the experts by taking only about three months to reach the required level. An area, regarded by many as the most magnificent example of classic Scottish Highland scenery, was thus transformed into an object of vast desolation and ugliness. Perhaps the most unsightly aspect is a wide sterile apron which girts the new loch's shore line, and upon which, because of the irregularity of the rise and fall, it is impossible for any life form to establish.

Viewed, objectively are such considerations only trivial and an affront to minority aesthetic sensibilities? Perhaps not. Figures recently produced dramatically illustrate the over capacity in electrical energy under *current* building projects. Ironically it is by virtue of this excessive capitalisation that electricity as a major power source increasingly inflates prices.

On the same projection, industry and the private consumer will, by the end of the century, pay most handsomely to cover the cost of this excessive investment. Were the Hydro schemes throughout the

Highlands really justified in the light of recent developments? Before he died old Kenny Mackay remarked to me "mind you boy, the light in the byre is handy but all the power in the world couldn't replace the pleasure of a fine day on Loch Monar".

Certainly the old style Lairds managed their Strathfarrar properties with a view to conserving natural amenities and maintaining a level of employment which ensured a degree of community life within the glen. The Stirlings showed themselves very conscious of sound principles in ensuring the wellbeing of their Monar Estate. In particular they paid attention to old proven practice for the improvement of the Red Deer breeding stock. The best stags were left to breed and not taken for trophies. A general policy of annually culling inferior stock to maintain equilibrium was their aim. In general, from the early eighteen hundreds until quite recently there seemed adequate revenue accruing outwith their agricultural or investment interests to allow many Proprietors to indulge such policies without too great a regard for the profitability of their deer forests.

The financial position of some estates hitherto operating along these traditional lines has now come under closer scrutiny. Gradually falling industrial wealth and profitability in U.K. has forced changed attitudes upon many Highland Landowners. Glen Strathfarrar did not escape these changes. None of the three Estates now runs a stock of hill ewes. Sheep no longer graze any of the glen's 30 mile length other than for a Keeper's flock which still remain on Monar. Shepherds' houses became holiday homes.

Interest in Red Deer in this and many other glens is turning to a more controlled form of management. At Culligran Estate much of the in-bye land denuded of cattle and sheep carries a deer farming enterprise. A herd of hinds held within high mesh fencing will be cropped for venison. Further up the glen the Lovat Estate, whilst not abandoning the glen for summering a large herd of hill cattle, are certainly paying heed to the venison producing potential of the glen and carefully feed their deer stocks through the winter. The hardiness of the deer, their excellent feed conversion ratio and easy tame-ability, make them seem an obvious choice for this new low cost venture in the hills. However a wider market for venison is still largely unpioneered and success in expanding the Red Deer's meat producing potential by more intensive methods is by no means assured. To its supposed advantage the taste of farmed venison is said to be less strong than that of the wild product.

It is pertinent to refer to the question of humane killing as deer farming expands. For the moment there is no restriction on deer killing other than those applied to the use of firearms, and the operation of a closed shooting season. Stalking is generally out of the public eye and kept in low profile by most of the Estate owners. Without doubt however a degree of poor shooting and wounding does take place using the stalking approach to killing but unlike fox hunting this has so far drawn little public comment. There may be some clash with sporting interests once deer farming becomes more widely established, as killing methods used on tame deer could come under closer scrutiny.

Recently we have seen the introduction of Red Deer trophy hunting. Sportsmen, invariably Germans, seem willing to pay high figures for the privilege of shooting a stag. This in some ways is not dissimilar to the great sporting tenants of a hundred years ago, though the sportsmen are treated rather more as holidaymakers and appear for a week rather than taking over the estate for a complete season. The Estate or some agent provide the necessary facilities and employees. Often the deer are enclosed by fencing and suitable hides erected. Ponies for deer carrying have now been largely replaced by small, ground hugging, broad wheeled, petrol engined vehicles. The shooter may well stay in a hotel which makes him suitable bookings for a day's sport. In short a more commercially orientated approach has been forced upon some Estate Proprietors which contrasts markedly with the management attitudes of pre-war days.

A number of Government bodies are increasingly involved in matters of land usage, ownership and control. So far the State's biggest land owning organisation, the Forestry Commission, has not acquired land in Strathfarrar, though much is eminently suitable for planting. This indicates the Proprietors unwillingness, until now, to sell the land. Although the Commission does not take up powers of compulsory purchase, its spread during the past forty years across vast areas of Scotland, previously under sheep or deer population, must again reflect the low return on this type of land.

By contrast the Countryside Commission and its offshoot the Nature Conservancy do have arbitrary powers at least of land designation. Accordingly, in the view of these bodies, a particular area's value in term of its flora and fauna is assessed and certain chosen places are termed Sites of Special Scientific Interest.

The Nature Conservancy is actively involved in the lower woodlands of Strathfarrar and the controlling of access to the glen.

Throughout U.K. these two bodies are beginning to feel their feet and move along a pathway of securing a greater influence in how the land will be treated and what forms of activity are best in keeping with principles they themselves have laid down.

Where lies the future for Glen Strathfarrar? Should it be designated a National Park as was mooted many years ago? Is there a case for greater public access? The road brought under Regional Authority jurisdiction and the glen opened to the full effects of the tourist industry? Conversely others might take the view that even greater restrictions upon access would ensure the area remained largely "wilderness". At least they would argue some portion of the Highlands could then exist as a form of wild life sanctuary untrammelled by an excessive human intrusion. Freedom could be afforded officials and scientists to study the glen's wealth of natural history untainted by the pollution of tourists intruding tents and plastic litter disturbing their studies.

One major trend over the past twenty years, is quite simply, fewer and fewer individuals or families now wish to live in a remote environment, or 'close to nature.' The weather is better viewed from behind double glazing, wild life more amenably studied during short forays from an office or laboratory. The call of the wilderness is being catered for adequately by the television documentary. Scenery looks best through a car window, you don't have to walk on it. The awesomeness of man challenging the elements is more comfortably experienced between the covers of a book.

Are the great masses wrong? Does an innate human requirement for the wild and primitive really exist? If so is its spirit being successfully checked and stifled? Man is not yet ready or capable of controlling the natural environment but his expanding ingenuity coupled with his desire to do so may yet prove his undoing. The countryside is not just a play area for picnickers, planners or scientists. It is part of our living self which responds by giving us, through our senses and emotions, a feeling of mutual care and belonging. We should embrace this harmony with a sense of affection and environmental morality.

The hills of Strathyre. JKT. 82

BURNING YESTERDAY

Diesel drench and lighted match
Undo by simple day
Cared labour's hand made years
Of cut stone skill
And shapely woods delight.

Spurting red the timbers blaze
Bursting blood of flame.
Through eyes of pain
Square sockets rage
Insane with ageless grief.

Crackling loud the timbers crash
With hideous licking hiss.
Thrown wide to sky
Once memoried walls
To tears of bitter rain.

Acrid smoke that chokes the blue
In purple progress robes,
Billowed bold they proudly drape
Man's naked power
To build and self destroy.

Last trailing plume from embered glow
Sends heated spits of stone
To greet a night
Which wraps one bone in seeing thought
To bare morn's laughs and scorn.

East sun shafts the sightless eyes,
Black rimmed, staring wild,
Straining westwards, pleading now,
One last gauntful gaze
Upon the comfort hills.

Charge of doom, the echoed roar
Explodes charred useless frame,
Whilst rumbled dust
Blinds watching eyes
With centuries' sad resign.